OUR ENDANGERED ATMOSPHERE

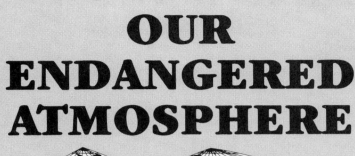

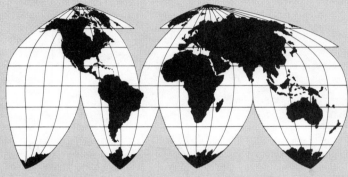

Global Warming & The Ozone Layer

Gary E. McCuen

IDEAS IN CONFLICT SERIES

publications inc.

502 Second Street
Hudson, Wisconsin 54016

Illustration & photo credits

Carol & Simpson 75, 89, 129, Jeff Danziger 119, EPA 11, 29, 34,
Info-Graphics 17, Republic of Panama 23, Bill Sanders 61, 128, Stayskal
101, USA Today 47, US Senate Committee on Environment and Public
Works 83

© 1987 by Gary E. McCuen Publications, Inc.
502 Second Street • Hudson, Wisconsin 54016
(715) 386-7113
International Standard Book Number 0-86596-063-1
Printed in the United States of America

CONTENTS

CHAPTER 3 THE OZONE LAYER

CHAPTER 4 THE NEED FOR ACTION?

REASONING SKILL DEVELOPMENT

These activities may be used as individualized study guides for students in libraries and resource centers or as discussion catalysts in small group and classroom discussions.

IDEAS in CONFLICT ®

This series features ideas in conflict on political, social and moral issues. It presents counterpoints, debates, opinions, commentary and analysis for use in libraries and classrooms. Each title in the series uses one or more of the following basic elements:

Introductions that present an issue overview giving historic background and/or a description of the controversy.

Counterpoints and debates carefully chosen from publications, books, and position papers on the political right and left to help librarians and teachers respond to requests that treatment of public issues be fair and balanced.

Symposiums and forums that go beyond debates that can polarize and oversimplify. These present commentary from across the political spectrum that reflect how complex issues attract many shades of opinion.

A global emphasis with foreign perspectives and surveys on various moral questions and political issues that will help readers to place subject matter in a less culture-bound and ethno-centric frame of reference. In an ever shrinking and interdependent world, understanding and cooperation are essential. Many issues are global in nature and can be effectively dealt with only by common efforts and international understanding.

Reasoning skill study guides and discussion activities provide ready made tools for helping with critical reading and evaluation of content. The guides and activities deal with one or more of the following:

RECOGNIZING AUTHOR'S POINT OF VIEW

INTERPRETING EDITORIAL CARTOONS

VALUES IN CONFLICT

WHAT IS EDITORIAL BIAS?

WHAT IS SEX BIAS?
WHAT IS POLITICAL BIAS?
WHAT IS ETHNOCENTRIC BIAS?
WHAT IS RACE BIAS?
WHAT IS RELIGIOUS BIAS?

*From across **the political spectrum** varied sources are presented for research projects and classroom discussions. Diverse opinions in the series come from magazines, newspapers, syndicated columnists, books, political speeches, foreign nations, and position papers by corporations and non-profit institutions.*

About the Editor

Gary E. McCuen is an editor and publisher of anthologies for public libraries and curriculum materials for schools. Over the past 16 years his publications of over 200 titles have specialized in social, moral and political conflict. They include books, pamphlets, cassettes, tabloids, filmstrips and simulation games, many of them designed from his curriculums during 11 years of teaching junior and senior high school social studies. At present he is the editor and publisher of the *Ideas in Conflict* series and the *Editorial Forum* series.

CHAPTER 1

OVERVIEW: THE GLOBAL ENVIRONMENT

OVERVIEW: THE GLOBAL ENVIRONMENT

WHAT IS THE GREENHOUSE EFFECT?

Carl Sagan

I would like to provide some assurance that there is such a thing as a greenhouse effect and that it can be serious, and at the end of my remarks to say something about the general question of constantly stumbling upon new potential catastrophes of climatic and other sorts, and the question of what, if anything, can be done at least to keep a little bit ahead in this sequence of potential catastrophes that seem to be discovered, one every few years.

If you imagine a planet which has no atmosphere and no clouds, what determines the surface temperature? Well, in principle there would be two sources of energy. There might be internal energy coming up from the inside of the planet—this is certainly true for Jupiter, for example, to a significant degree—and there is energy coming from the outside, and that is almost entirely from the Sun.

In the case of the Earth and the inner planets, there is no significant energy coming from the inside, so the source of the temperature of the surface of the planet is sunlight. If there were no atmosphere, then how does the amount of sunlight determine the temperature? Well, if the planet is highly reflective, then less sunlight is absorbed and goes into heating the place.

Excerpted from testimony by Carl Sagan before the House Subcommittee on Investigations and Oversight, February 28, 1984.

If a planet is not very reflective, if it absorbs a lot of sunlight, then that light goes into making the place hotter.

This reflectivity, the ability to reflect light back to space, is often called the albedo and it can vary. Freshly fallen snow, for example, has an albedo of maybe 70 percent, and black velvet has an albedo of a few percent, and most things fall in between. Deserts are 20 percent, and so on.

Now if the Earth had its present albedo but had no atmosphere, the average temperature of the Earth would be well below the freezing point of water. This planet would, in fact, be uninhabitable if there were no greenhouse effect. It is the greenhouse effect that brings the average temperature of the Earth well above the freezing point and permits oceans and bodies of water and all that.

Greenhouse Effect

So now let's say something about greenhouse effect. In a very simple way, what happens in the greenhouse effect is something like this. First of all, let me say it is a misnomer, the phrase "greenhouse effect," in that greenhouses don't work that way. But this is a historical error we needn't go into. Now consider a planet with air, like our own, which is clearly transparent—except in Los Angeles—to sunlight. So visible light from the Sun comes through the atmosphere and strikes the surface and, as we were saying a moment ago, warms it up. The surface, like every other object in the universe that is not at absolute zero, radiates and at the temperatures of the surface of the Earth, it radiates mainly in the infrared part of the spectrum, radiation that is longer wave than the red part of the visible spectrum. You do not directly sense infrared radiation, sometimes called heat radiation, but it sure is there.

What happens is, a kind of equilibrium is established so that the amount of sunlight coming in from the Sun that is absorbed by the planet is precisely balanced by the amount of infrared radiation emitted by the planet back to space. Now, in the case of a greenhouse effect, the visible light comes in just as it would have if there were no atmosphere, but the atmospheric gases invisible in the visible—that is, transparent in the visible part of the spectrum—tend to be opaque in the infrared part of the spectrum. The thermal radiation in the infrared is impeded

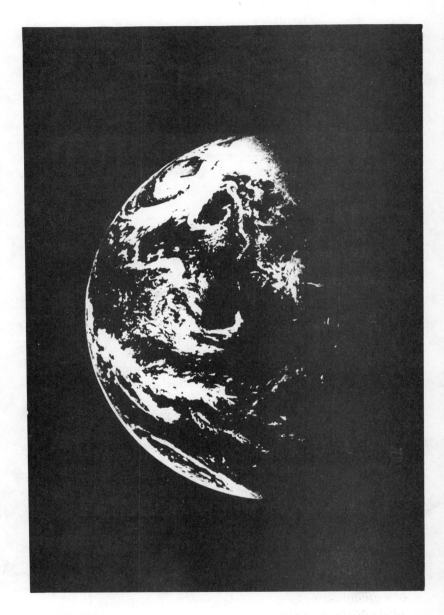

Source: EPA

from getting out. You can consider it as a kind of blanketing of the Earth in the infrared and not in the visible part of the spectrum.

As a result, the surface temperature has to go up until the radiation which is leaking out in the infrared where there isn't a lot of opacity just balances the visible radiation that is coming in. And such greenhouse effects can be very significant. In the Earth's atmosphere, the greenhouse effect is due mainly to carbon dioxide and to water vapor, with other constituents playing more minor roles, although oxides of nitrogen and other materials could be significant on a smaller scale. . . .

Mutual Agreement

Now, the scientific community is in very good mutual agreement on the overall general consequences of the burning of fossil fuels, putting more carbon dioxide into the atmosphere of the Earth, producing an incremental greenhouse effect that is adding to the existing greenhouse effect. The greenhouse effect, I stress again, is a good thing. We owe our lives to it. And while differences between calculations have been stressed in the press, what is important is that all the calculations agree to the first order that doubling of the carbon dioxide in the Earth's atmosphere will increase the global temperature by a few degrees Centigrade, something of that order.

To require that scientists provide an absolutely ironclad, guaranteed value of how much the temperature will go up is probably asking too much. The calculations involve many factors, and you cannot be absolutely sure that you have included every one of them. What is striking is the unanimity of all of the calculations, so if a few degree increment in the global temperature is a bad thing then you ought to start worrying about what to do in that case. Also, you ought to start worrying about whether there is some way to avoid putting more carbon dioxide into the atmosphere.

These are questions of cost effectiveness. It is not a catastrophe like, for example, nuclear winter would be. It is slow, a period of many decades or a century before the effect we're talking about is fully manifested, and presumably even the worst of it is not as bad as lots of other things we could think about. But it does raise worrisome general questions. Our technology is now able to make significant changes in the environment of the planet we live in, and who would have thought that burning

Astronauts Say Pollution Shocking

The astronauts of space shuttle Challenger said Friday they were shocked during their five-day flight at the amount of pollution they saw surrounding the earth below them.

"It was appalling to me to see how dirty our atmosphere is getting," Paul J. Weitz, commander of the first flight of the Challenger, said at a press conference at Houston Space Center.

"Unfortunately, this world is rapidly becoming a gray planet," Weitz said. "Our environment apparently is flat going downhill . . . What's the message? We are fouling our own nest."

Weitz said the heavy pall of pollution resembled the blue smoke of burning wood and seemed to hang over all the world's land masses, even undeveloped countries.

Associated Press, April 23, 1983

wood and coal to keep warm would have this quite unexpected consequence of keeping everybody warmer than they were?

When you run through the full range of consequences of modern technology, including industrial pollution, including the possible consequences of nuclear war, including the on-again, off-again concern about halocarbon propellants in spray cans, there is an overall impression, which is that we are pushing and pulling on the global environment because of innocently intended high technology, in ways that we do not fully understand and in ways that may have serious consequences. That naturally raises the question of, what institutions are there whose job it is to try to detect, identify, and if possible defuse such problems early enough.

If you look at any of these issues you find that the problems have been identified by individual scientists in the academic community, mainly who were worrying about something else and happened to stumble upon this or that effect. That got them worried and they talked to their colleagues and got other people worried. There is no institution whose job it is to systematically seek out such effects.

If we have found a half a dozen such effects in the last 20 years, let's say, it is likely, it seems to me, that there are some more that we haven't stumbled on yet, and for all we know they are more serious than any of the ones we're talking about yet, except for nuclear winter. It's hard to imagine something more serious than that. Is there some institutional framework in which these problems can be systematically sought out and addressed? It seems to me an important problem.

OVERVIEW: THE GLOBAL ENVIRONMENT

CARBON DIOXIDE
AND GLOBAL WARMING

John R. Justice

Might man's releases of carbon dioxide (CO_2) change the global climate and environment? The possibility of CO_2-induced climatic change has been at the forefront of thinking, writing, and speaking for many years. The question was raised at the end of the last century, crystallized 25 years ago during the International Geophysical Year, and politicized during the last decade. It has elicited increasing amounts of interested discussion and careful analysis in the scientific literature, as well as speculation in the public press with talk of cornucopias and catastrophes, of winners and losers, and of greenhouses and ice ages. CO_2-induced climatic change has become a focus of newspaper editorials, congressional hearings, and an international research program involving hundreds of scientists.

Despite a large number of remaining uncertainties, the importance of the CO_2 issue has been underscored by a growing consensus of climate and atmospheric scientists in academia, government, and international scientific organizations. According to various predictions, the increased CO_2 might warm the Earth's

Excerpted from a report of the House Committee on Science and Technology, October, 1984. John R. Justice is a specialist in Earth and Ocean Science.

climate, change precipitation and temperature regimes regionally and seasonally, and alter large-scale oceanic and atmospheric circulation patterns. Scientists continue, however, to debate *when* the Earth will respond and *what* impact this might have on mankind. . . .

What is the Greenhouse Effect

The carbon dioxide molecule plays an important role in modifying the vertical distribution of temperature in the atmosphere by absorbing infrared radiation. The larger the concentration of atmospheric carbon dioxide, the more opaque the atmosphere is to infrared radiation emitted by the Earth's surface. The more outgoing infrared radiation trapped in the atmosphere, the higher the air temperature becomes. This absorption by the CO_2 molecule of infrared radiation and subsequent elevation of air temperatures near the surface of the Earth is known as the greenhouse effect, although somewhat misnamed. (Actually, a greenhouse warms its environment largely by reducing thermal losses due to convection. The warming associated with a CO_2 greenhouse effect can be attributed to reducing the amount of outgoing infrared radiation lost to outer space, not to reducing convection losses.)

Sources and Sinks for Carbon Dioxide

During the last century and a half, and particularly in the post World War II period, global use of fossil fuels has grown markedly, and the burning of these fuels is believed to have been the principal source of increasing atmospheric CO_2. At current rates of fossil fuel consumption, approximately 5 billion tons of carbon (as CO_2) are released annually to the atmosphere. An important factor, however, is the amount of this CO_2 that remains in the atmosphere—what the scientists refer to as the airborne fraction. About half of the total amount of CO_2 released from fossil fuels remains in the atmosphere, resulting in the measured increase in concentration. Scientists have assumed that the world ocean and both the land and marine biospheres serve as reservoirs (sinks) for the other half of this excess CO_2, but the fate of this missing 50 percent is the subject of active research. . . .

16

The "greenhouse" effect

Recent studies have renewed interest in the "greenhouse" effect, caused by increased levels of carbon dioxide and water vapor in the atmosphere. Carbon dioxide is released when fuels are burnt and it is absorbed by growing plants. Two factors are increasing the amount of CO^2 — the burning of fossil fuels, producing CO^2 and the felling of the tropical forests, resulting in less absorption.

The extra carbon dioxide is forming a layer in the atmosphere like glass in a greenhouse — allowing heat from the sun in, but not letting it escape again. The heating process is accelerated by the rising temperature causing more evaporation. This in turn adds more water vapor to the atmosphere, helping to trap more heat.

Eventually, experts predict the rise in average temperatures will mean reduced crop yields in the wheat and corn belts, water supply problems in the dryer states and a higher sea level from melted polar ice.

Carbon dioxide from burnt fuel

Greenhouse layers

Water vapor in atmosphere

Carbon dioxide in atmosphere

Heat from sun reflected back

More evaporation

Felled forests absorb less CO^2

Farm production affected

Higher ocean level

Data/EPA, Carbon Dioxide Assessment Committee

Source/InfoGraphics

Info-Graphics © by and permission of News America Syndicate

Temperature Effects

Will the increase in CO_2 cause a significant rise in globally averaged surface air temperatures, or might some currently known or some yet unknown countervailing climatic factor or feedback effect moderate the warming trend or even reverse it? If today's most widely accepted modelling results are correct, a doubling of the preindustrial atmospheric CO_2 concentration (i.e., from

300 ppm to 600 ppm) could increase the average temperature of the lower atmosphere at middle latitudes by about 3° Celsius.

The Detection Problem: Finding the Signal in the Noise

Efforts to detect, through actual observation, an unambiguous signal of the CO_2-induced climate response predicted by current models are important for both scientific and policy aspects of the carbon dioxide question. First, current conventional wisdom holds that nothing will be done about CO_2 until, at a minimum, a clear signal of actual warming is detected. Regardless of how warranted this view may prove, it seems evident that unambiguous detection of the predicted changes would affect the terms and credibility of the CO_2 debate. Second, detection studies are an important means of confirming, rejecting, or modifying models of the interaction between atmospheric CO_2 and global climate.

Current observational evidence does not prove that a CO_2-induced climatic warming is occurring. On the other hand, it should be emphasized that the evidence does not allow rejection of CO_2-climate theories either.

Effects of Increasing CO_2 on People and the Environment

To begin with, it is important to put the climate change that could be induced by carbon dioxide into realistic perspective. Increasing CO_2 may hardly be the major challenge facing humanity. Instead, it might reasonably be viewed as one of many chips that may figure in the future balance of the affairs of nations. CO_2 effects may tip some scales, but this will depend largely on where other, perhaps bigger chips happen to fall. There is growing recognition that a useful evaluation of CO_2 effects must weigh the influences of population growth, energy and economic development, and technical change as well.

The most important effects of CO_2-induced climate change will probably involve the supplies of food, fiber, and water that people seek to obtain from their environments. A longer term and probably less urgent concern is the possible effect of CO_2 on human settlement patterns. But both local climate change and a global rise in sea level, if large enough, could eventually affect where and how future populations can live.

A CO_2 buildup implies large economic, social, and political impacts, if the model-stimulated climate responses are correct, but not necessarily disaster. In fact the direct impact on crops

18

and forest growth may be positive, allowing for technological adaptation. The common prejudgement that all environmental change is bad should be avoided. The key question is: What actions, taken when, would help to maximize the opportunities and minimize the dangers implicit in the CO_2 issue? How could food, fiber, and water systems be made more adaptive to the kinds of stresses and opportunities that a changing climate would entail? What makes these systems vulnerable to climatic disruption? . . .

Summary and Conclusions

The increase in the concentration of CO_2 in the atmosphere and the possibility of CO_2-induced climatic change have become subjects of increasing interest in the media as well as in the scientific community. The buildup of CO_2 is a reality, monitored with increasing precision since 1957 and inferred from much earlier dates. Indeed, about the only facts available are the actual measurements of CO_2 concentration, particularly the two-decade series at Mauna Loa and the South Pole, and some fairly reliable data from the United Nations on the annual consumption of fossil fuels in industrial countries. There the confidence ends. The CO_2 "question" is actually several questions, characterized by a litany of unknowns and uncertainties—of the internal transport processes in the world ocean, of the nature and timing of the climatic response, of first-order detection, of the magnitude of forest and soil carbon wastage, and of the future course of fossil fuel consumption. Yet something else emerges, too: *if* the CO_2 buildup continues, *if* the big general circulation models are right about its impact on climate, and *if* the potential role of the ocean and the land and marine biospheres has not been miscalculated, *then* there is the possibility of a climate change in the next century and a half like nothing the post glacial world, and hence civilized humanity, has seen.

The scientific community has learned enough to run up warning signals in certain cases, but not enough to quantify the dangers with adequate confidence. So, what can climate and atmospheric scientists say now? . . .

The uncertainties inherent in the CO_2 problem are often cited as justification for taking no action now. But, whether CO_2 builds up or not, it is known that weather and climate will continue to fluctuate. Dealing with the CO_2-climate problem is one way of coping with climatic variability in general. Thus, there are incentives to reduce the vulnerability of human settlements and activities to both climatic change and variability, to begin actions to mitigate adverse CO_2-induced impacts, and to take advantage of the beneficial effects. Table 1 lists some specific strategies that could help to increase resilience and mitigate the effects of global climate change. They would likely be advantageous even if no climate change were to occur. Moreover, some of these strategies might also slow the increase in atmospheric carbon dioxide.

TABLE 1. Strategies To Mitigate the Effects of Increased CO2 or To Help Avert the Climatic Change

Strategies that:

Increase resilience	Reduce CO2 emmissions	Improve choices
Protect arable soil	Energy conservation	Environmental monitoring and warning
Improve water management	Renewable energy resources	Provision of improved climate data and its application
Apply agrotechnology	Reforestation	Public information and education
Coastal land use policies	Energy taxes on use of fossil fuels	Transfer appropriate technology
Maintain global food reserves	Remove CO2 from flue gases after combustion	
Disaster relief	Weather and climate modification schemes	

OVERVIEW: THE GLOBAL ENVIRONMENT

TROPICAL FORESTS AND THE ATMOSPHERE

José A. Lutzenberger

We are witnessing today in Brazil and in much of Latin America the biggest holocaust in the history of life. Never in the course of three and a half thousand million years, since the first stirrings of life on this planet, has there been such a wholesale, accelerated, violent and irreversible demolition of all living systems as today. We have passed the point where we only desecrate this or that scenic landscape, this or that ecosystem. We are now in the process of demolishing whole biomes. Now we are getting ready to finish off the last large, more or less intact and contiguous jungle on Earth, the Hylaea, or tropical rain forest in Amazonia.

Nobody seems to know exactly how fast the forest is being demolished. No statistics are published and landstat images are not easily accessible to and interpreted by those who really care, while the powerful have an interest in downplaying the extent of the devastation. In a recent interview on TV the Governor of the State of Amazonas, who is now pleading for free, largescale export of raw timber logs and animal hides, stated categorically that less than 2% of the forest had been cleared until now. On the other hand, serious scientists who know Amazonia very well, Harald Sioli, f.i., state that close to 100,000 km^2 of forest are

Excerpted from a written statement before the House Committee on Science and Technology. José A. Lutzenberger is a Brazilian scholar who has studied the devastation of the Amazon tropical forest.

felled every year. That would be about 2% a year and this has been going on for at least a decade. . . .

The systematic destruction is done in the name of "progress." The Brazilian Government, the military dictatorship which set itself up in 1964 set course for "development" at any cost.

Large scale devastation of the tropical rain forest and its surrounding transition type forests takes several forms. At one extreme we have gigantic projects. Multinational or Brazilian corporations or powerful individuals go to Amazonia to multiply capital. Among them are such giants as Anderson Clayton, Goodyear, Volkswagen, Nixdorf Computer, Nestle, Liquigas, Borden, Kennecot Copper, and the American multibillionair David Ludwig, or even farmers cooperatives from the list. It runs into the hundreds. These outfits set up enormous projects—cattle ranches, paper mills, single species monocultures of exotic trees for pulp, immense rice plantations, sugar cane plantations for the alcohol program, timber mills, mining operations. . . .

Devastation

We must remind ourselves that the intact forest, obliterated to give way to pasture, can produce at least ten times as much food in the form of tropical fruit, game and fish. Every single adult Brazil nut tree left in peace can produce hundreds of kilos of precious food, every pupunha palm tree or many of the innumerable other species of palm trees occurring in the forest can produce dozens of kilos of food, feed and construction material. For the inhabitants of the forest there also is no limitation concerning firewood, a problem that is becoming extremely serious in other parts of the world.

Another devastating effect of those schemes, this one social, is that they employ an average of one worker per two thousand cattle, that is, one person on at least 3000 hectares! The same area of forest could easily feed and house several hundred people if left intact. The traditional life style of the Indian, the caboclo and the seringueiro (rubber tapper) is also much more pleasant, easier, independent, and secure than the life style of the ranch worker. The irony of it all is that the little meat produced is meant for export. The Amazonian caboclo wisely says—where cattle move in, we move out, cattle mean hunger. The only beneficiaries are the corporations who don't even spend the money in the areas they devastate. But they keep saying that they are in the business of feeding starving humanity. . .

FORESTS OF PANAMA – 1983

Forested areas

Source: Republic of Panama

The social devastation of the other schemes, extensive monoculture of trees, open pit mining, gigantic dams, timber mills and logging on an industrial scale, commercial fishing for export, are just as bad. They are all geared to the enrichment of the powerful from outside the region. There is no concern for the needs of the local population, much less for their life style and culture. The local people are uprooted, marginalized, alienated and they go either to the slums or escape ever deeper into the jungle, as long as there is jungle. The Indians are already reaching the end of the line.

The Brazilian government is now selling off whole mountains, such as in the Carajá Project. Recently our minister of planning boasted of having received the first downpayment of a few hundred million dollars from Japan for ore to be mined in the Carajá mountains. What will future generations say? . . .

Amazonia

Amazonia should be left to the Amazonians, it is theirs. The growth of capital and power at the expense of the ecology and the people of Amazonia is classical imperialism. It makes no difference whether the benefits accrue to powers from overseas or from other parts of Brazil.

There is one more issue that must not be forgotten. Studies such as those of Salatti (University of Piracicaba) have shown that the climate is largely generated by the forest itself. Evapotranspiration is so intense that 50% or more of rain water is put back into the atmosphere. The rain that falls on the slopes of the Andes is water that has been recycled between 5 and 7 times on its way from the Atlantic. The whole system of recycling may break down if the chain is interrupted at the start. But exactly this is being done today. The forest of Pará may be gone by the year 1990. Already now we have an exceptional drought on the Island of Marajó in the mouth of the AMAZON River. Where the forest disappears, intensive evapotranspiration gives place to naked soils that heat up to 50° C or more in the hot tropical sun producing hot upwinds that dissolve clouds instead of producing them as does the forest. The rainfall pattern may change all along the way to the Andes. The rain forest, with its shallow root system, cannot stand long periods of drought. In this respect it is fundamentally different from the savannah (cerrado) forests farther south where vegetation has very deep root systems. Before it dies, the rain forest may

Rain Forest Destruction

In tropical lesser-developed countries—like Indonesia, Brazil, and the nations of Central America—the rapid destruction of the world's tropical rain forests is rapidly creating an alarming environmental and economic disaster. In the name of "development," many of these nations are cutting down forests so quickly and recklessly that, by the time people might be roused to the danger and take action to stop it, it could be too late. . . .

Furthermore, a devastating heating up of the Earth's surface will likely result from the rain forests' destruction. Evaporation and other processes that take place in the forests serve to moderate global temperatures and air currents. And the tremendous amounts of carbon dioxide now trapped within the rain forests, would, once released, intensify the "greenhouse effect" that is already being worsened by industrial pollutants.

The People, September 28, 1985

become combustible. Already around Belém a few weeks of drought give rise to brush fires. The first fires may not kill the forest, but successive fires over a few years will do it. Brazilian pyromania is proverbial. A process of successive demolition may set in.

I have already observed another type of self-demolition on some of the lower areas of flood plains. Some of the igapós (flood plain forest) are dying without being touched by chainsaw or defoliant. Deforestation in the upper Amazon basin, such as in Rondonia and Peru unbalances the flow of the river. Floods are higher than normal, drought is more prolonged. The Igapó community gets more and less than it can stand.

World Climate

Collapse of the rainforest or large scale deforestation will inevitably lead to changes in regional climate. It is difficult to see how a serious change in regional climate will not affect world climate. The Amazon rainforest is evenly spread over both hemispheres. The fresh water that flows and is recycled through this well balanced system constitutes about one quarter of total world river flow. Together with the carbon dioxide balance, the ozone layer, aerosols, dusts, cloudiness and albedo, all of which are being systematically and blindly disturbed by modern industrial society, the Amazon rainforest is one of the important regulatory mechanisms of world climate.

Those of us who live in temperate or subtropical climates should perhaps be the most concerned. A change in world climate, even small, will affect us more than the tropics. During the ice ages or inter-glacial periods temperatures on the equator did not change very much but the tropical belt shrank or broadened, the subtropical and temperate belts moved towards or away from the equator. It is easy to imagine what this will mean for Europe, Asia, the US, southern Africa, southern South America, and Australia.

In our fight for ecological sanity we must always keep in mind that social justice and a healthy environment go together. Humanity will only abandon its present suicidal course when the masses become ecologically conscious and exert enough pressure on the powerful, whether they be multinationals, local oligarchies or governments of whatever denomination.

Brazil's policies concerning the Amazon must change, they must change during this decade, or it may be too late!

OVERVIEW: THE GLOBAL ENVIRONMENT

THE OCEANS' EFFECT ON CLIMATE

Carl Wunsch

I will discuss the relationship between the ocean and the continuing rise in levels of atmospheric carbon dioxide (CO_2). That the ocean is important in understanding the effects of the CO_2 rise has been known for a long time. But the extent to which the oceanic effects introduce very great uncertainty has not been much appreciated.

Rising concentrations of carbon dioxide and other trace gases with "greenhouse" properties are now widely recognized as representing an "experiment" with our environment with potentially catastrophic consequences in the longer term. The weight of scientific understanding strongly supports the conclusion that the atmosphere will warm considerably (an average change of about 2° C, with regional changes probably much larger than this average), and with a greater warming near the poles than at the equator. Although no one can guarantee this outcome, it is widely accepted as the most likely change.

Role of the Ocean

What is the role of the ocean in the actual result? In general terms, the ocean plays a number of parts. Much of the carbon

Excerpted from testimony before the Senate Committee on Environment and Public Works. Carl Wunsch is a professor of physical oceanography at the Massachusetts Institute of Technology, Cambridge, Massachusetts.

dioxide that is added to the atmosphere ultimately finds its way into the ocean. To the extent that CO_2 is dissolved in the *deep* ocean, it no longer has a greenhouse effect and thus reduces the effective atmospheric warming. The fraction of the CO_2 that will eventually be locked up in the ocean can be calculated on the basis of chemical equilibrium and is not controversial. Unfortunately, the time required to reach this final equilibrium state is extremely long (probably thousands of years) and so it is not particularly relevant to the present discussion. . . .

The ocean thus removes part of the carbon dioxide and the rate at which it does so in the near-term will to a great extent determine the rate at which the atmosphere warms. Should the ocean remove a great deal more than is currently estimated, the warming will be delayed; should it take up a good deal less, the warming will come sooner.

Less Direct Roles

But the ocean plays a number of less direct roles whose consequences also need to be understood. As the atmosphere warms, it in turn warms the ocean. Not only is the ocean a sink for CO_2, it is also a sink for the increasing heat in the atmosphere. When heat is placed in the ocean rather than accumulating in the atmosphere, it takes longer for the atmosphere to warm up to the level that the greenhouse effect will eventually cause.

The ocean, as it absorbs heat in some locations, moves that heat around as the fluid circulates, giving it back to the atmosphere in other locations (the heat put into the ocean in the tropics and carried to mid-latitudes where it is returned to the atmosphere is largely what makes the mid-latitudes habitable for us). Thus the regional changes in warming, cooling, and precipitation patterns that will be caused by the warming, will be partly determined by how the ocean moves the increased heat around. Recall that the extremely intense effect El Nino has upon our weather and climate is the consequence of a change in ocean surface temperatures in the tropics of no more than about 1° C.

What Will Happen?

Some conclusions about what will happen are reasonably clear. There will be shifts in climate, both locally and globally, with probably large shifts in rainfall patterns. Sea level will rise as

GLOBAL TEMPERATURES AND SEA LEVEL
RISE IN THE LAST CENTURY

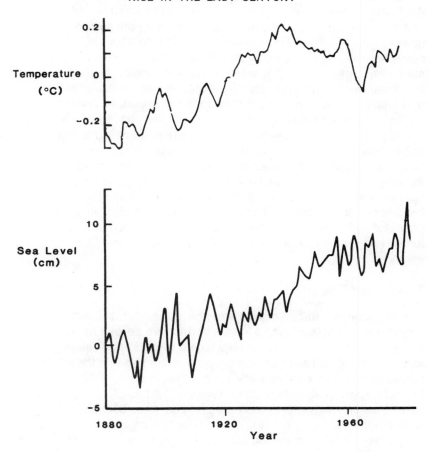

Source: EPA

the ocean warms, causing potentially serious flooding problems
(the rise in sea level will occur both because of the melting of
ice, but also because warmed water expands).

What is quite unclear is *how fast* these things will occur.
Society can deal with the consequences of such shifts on a 200
year time scale in a very different way than it can accommodate
them on a 20-50 year time scale. I believe it would be a foolhardy
scientist who would categorically predict which we are going
to see.

Much of the difficulty in making a time scale forecast lies in the nearly complete uncertainty over how the ocean itself will change as a warming proceeds. The ocean is a very complex, flowing, fluid machine. Because we know so little about how it operates, most of the calculations that have been done to understand the impact of the CO_2 warming have effectively treated the ocean as though it were a solid—able to absorb CO_2, heat and moisture and to give them back to the atmosphere, but not itself changing any of its existing dynamical (i.e. flow) properties.

From an oceanographer's point of view, the worry is that the ocean itself is going to respond in ways which are not accounted for in existing ocean models, and which can substantially change the rates and regions in which the CO_2 warming effects will appear. The geological record strongly suggests that such changes have occurred in the past. To give some of the flavor of the possibilities let me try to describe how we think the ocean works today, and *one way* in which it *could* change under the CO_2 warming. . . .

The Ocean Today

The ocean is a turbulent fluid and I have tried to represent only those gross, large-scale features which we believe have the most immediate impact on our climate system. The ocean waters which gain heat from the sun and atmosphere in the tropics tend to move on average toward the polar regions. In the polar regions, the atmosphere regains that heat, leaving behind water which is extremely cold and salty, so cold and salty that it tends to sink to the ocean floor in a process we call "convection." This sinking process is exceedingly important: it is the sinking which sucks the warm water up from the south to replace the water which has gone to great depths. One of the important reasons that western Europe and the US west coast are so warm is that the ocean gets so cold to the north of them, in a process leading to convection, which brings warm water flowing in from the south. Anything which changes the rate at which convection occurs will change the global climate.

Furthermore, water which is in contact with the atmosphere, i.e. the warm water being sucked up from the south, becomes saturated in CO_2. Once it is saturated, it can take no more out of the atmosphere. But when the saturated water sinks, it injects CO_2 into the deep ocean, thus removing it from the atmosphere and bringing unsaturated water to the surface and speeding up the CO_2 removal from the atmosphere. Any process which changes the speed with which the ocean convects, will affect the rate of CO_2 buildup in the atmosphere.

In addition, the water which moves poleward along and near the surface is in the region of the active biological productivity. Plant and animal production in these near-surface waters depletes it of the nutrients essential for biological activity. The cycle is maintained when water which has sunk at high latitudes in the convective process is eventually, and many years later, returned to near-surface, enriched in nutrients. Anything which reduces the convection rate, ultimately reduces the depth to which the water descends, the rate at which water returns to the surface, and hence, ultimately, the supply of nutrients to near-surface life.

What Could Happen in a Warming

We suppose that the near-surface layers of the ocean become somewhat warmer, at least initially and that the temperature increase in both the atmosphere and ocean are somewhat greater near the poles. We expect that the wind-driving on the ocean surface will be reduced, because the extensive wind systems (trades and westerlies) are largely due to the great temperature contrast between high and low latitudes. With the wind driving reduced, the rates at which the ocean water moves will in turn be reduced. The combination of decreased wind driving, and somewhat warmer near surface waters and air temperatures, implies that water now sinking at high latitudes would not get as heavy as today, and therefore would not sink as fast, or as deep, thus further reducing the rate of ocean circulation. The ocean would be able to take up less heat than it now does (unable then to push it downward) and less CO_2 (also unable to push it downward). The CO_2 content of the atmosphere would rise more rapidly than now envisioned. . . .

What I have outlined is of course speculative, but it is possible.

OVERVIEW: THE GLOBAL ENVIRONMENT

SEA LEVEL RISE AND COASTAL LIVING PATTERNS

Stephen P. Leatherman

Throughout geologic history sea level has fluctuated greatly. During the last ice age (approximately 15,000 years ago), sea level was as much as 100 meters below present levels. The earth at this time was about five degrees celsius colder than today. During warm interglacial periods, sea level has been at times several meters higher than present. Because of the historic relationship between climate and sea level position, it is expected that anthropogenic (human-induced) global warming could cause a significant rise in sea level. Warmer temperatures could expand ocean waters, melt glaciers, and eventually cause the disintegration of the West Antarctic ice sheet. . . .

Future Sea-Level Rise

Concern about a possible acceleration in the rate of sea-level rise stems from measurements showing that concentrations of carbon dioxide and other "greenhouse" gases produced by human activities are increasing in the atmosphere. Because these gases absorb (trap) long-wave radiation (heat) in the atmosphere, it is generally expected that the earth will warm substantially in the

Excerpted from testimony before the Senate Committee on Environment and Public Works, June 10, 1986. Dr. Stephen P. Leatherman is an associate professor in the Department of Geography at the University of Maryland.

future. The National Academy of Sciences has convened two
panels to review all the evidence and concluded that warming
will take place. . . .

There is no doubt that the concentration of greenhouse gases
is increasing and will do so in the foreseeable future. However,
considerable uncertainty exists regarding the amount of warming;
it is generally agreed that a doubling of the greenhouse gases
will raise the earth's average surface temperature by about 1°
C if nothing else changed. It appears that most of the climatic
factors will amplify the direct effects, but some negative feedbacks
(such as increased cloud cover to offset part of the warming)
cannot be ruled out. Nevertheless, two panels of the National
Academy of Sciences (NAS) have concluded that a doubling of
greenhouse gases will eventually induce a warming between
1.5° and 4.5° C (3° - 8° F). . . .

Effects of Sea-Level Rise

The principal effects of sea-level rise are increased tidal
flooding and wave-induced erosion. Salt-water intrusion can also
be a problem in some areas, particularly affecting surface waters.

Tidal Flooding

A rise in sea level represents a raising of the water base
level. Therefore, storm waves and surges can reach higher and
further inland. This can result in accelerated beach erosion as
explained later, and major flooding will occur more often. For
example, "100-year" storms can occur on a 10 year averaged
basis by virtue of higher base levels when considering frequency-
magnitude relationships of coastal flooding.

The most significant impact of higher sea-levels will be the
submergence of coastal wetlands. Intertidal salt marshes can
adapt only to relatively moderate rates of sea level rise; rapid
increases in sea level can literally drown these wetlands, converting
them to shallow bodies of open water. . . .

A one-meter rise in sea level could drown most of the wetlands
without necessarily creating new marshes inland. Even in natural
areas the marshes will contract because of the sloping nature
of the land above the marsh plain. Where marshes are backed
by urbanized areas, such as along much of the Long Island,
N.Y. coast for example, these habitats will be squeezed out with
future sea-level rise.

33

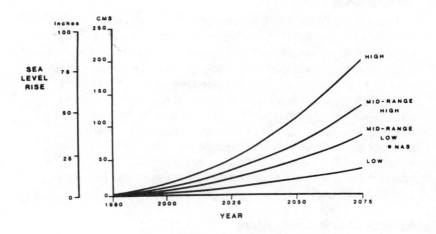

GLOBAL SEA LEVEL RISE SCENARIOS:
LOW, MID-RANGE LOW, MID-RANGE HIGH, AND HIGH

Source:EPA

Coastal Erosion

Sea level is one of the principal determinants of shoreline position. There are several reasons why sea-level rise would induce beach erosion or accelerate on-going shore retreat: (1) waves can get closer to shore before dissipating their energy by breaking, (2) deeper water decreases wave refraction and thus increases the capacity for longshore transport, and (3) with a higher water level, the wave and current erosion processes are acting further up the beach profile, causing a readjustment of that profile. . . .

Recommendations

1. Shoreline data from historical maps, charts, and aerial photographs needs to be compared by state-of-the-art mapping techniques to yield rates of beach recession. This information on a national basis is needed to provide a reference for and calibration of projected erosion based on accelerated sea-level rise. At present accurate historical shoreline data exist for only parts of the U.S. coast.
2. Salt marshes are already being lost at alarming rates in some coastal areas (notably in Louisiana and the Blackwater

Wildlife Refuge in Maryland). More research is needed to understand the mechanisms of marsh loss, which probably vary on a geographic basis.

3. An assessment of coastal urbanization along eroding shorelines needs to be made. What are the (national) trends in terms of continued development and increased vulnerability? How should coastal planning be modified to take into account accelerated sea-level rise?

4. Shore erosion and wetlands loss in association with coastal urbanization are critical research priorities. Academic scientists, who have been largely responsible for our present understanding of these processes, need to receive extended and expanded research monies to provide independent and objective data. In practical terms, this university research initiative should be *supplemented* by the relevant federal agencies, including the Corps of Engineers, Fish & Wildlife Service, EPA, and HUD.

WHAT IS
EDITORIAL BIAS?

This activity may be used as an individualized study guide for students in libraries and resource centers or as a discussion catalyst in small group and classroom discussions.

The capacity to recognize an author's point of view is an essential reading skill. The skill to read with insight and understanding involves the ability to detect different kinds of opinions or bias. Sex bias, race bias, ethnocentric bias, political bias and religious bias are five basic kinds of opinions expressed in editorials and all literature that attempts to persuade. They are briefly defined in the glossary below.

5 Kinds of Editorial Bias

sex bias— the expression of dislike for and/or feeling of superiority over the opposite sex or a particular sexual minority

race bias— the expression of dislike for and/or feeling of superiority over a racial group

ethnocentric bias— the expression of a belief that one's own group, race, religion, culture or nation is superior. Ethnocentric persons judge others by their own standards and values.

political bias— the expression of political opinions and attitudes about domestic or foreign affairs

religious bias— the expression of a religious belief or attitude

Guidelines

1. From the readings in chapter one, locate five sentences that provide examples of editorial opinion or bias.

2. Write down each of the above sentences and determine what kind of bias each sentence represents. Is it sex bias, race bias, ethnocentric bias, political bias or religious bias?

3. Make up one sentence statements that would be an example of each of the following: *sex bias, race bias, ethnocentric bias, political bias,* and *religious bias.*

4. See if you can locate five sentences that are factual statements from the readings in chapter one.

CHAPTER 2

GLOBAL WARMING: IDEAS IN CONFLICT

GLOBAL WARMING:
IDEAS IN CONFLICT

FACTS FAIL TO
BACK WARMING THEORY

Patrick J. Michaels

Patrick J. Michaels is the Virginia state climatologist and an associate professor of environmental sciences at the University of Virginia.

Points to Consider

1. How has temperature changed in the northern and southern hemisphere?
2. Why have the temperature changes in the two hemispheres gone in opposite directions?
3. Why is there little evidence for a global warming?

Patrick J. Michaels, "Facts Fail to Back Predictions About Global Warming," *Minneapolis Star and Tribune*, July 17, 1986. Reprinted with permission of the author.

The northern latitudes, from 24 to 90 degrees north, have actually cooled off since 1930.

Why do people who express so much skepticism at the daily weather forecast readily panic at one for 50 years from now?

The latest version, presented to a Senate committee last week by NASA, left the impression after the media got through with it of unqualified gloom and doom by 2030.

It's one thing to say that a disastrous global warming could happen. But no computer model, as noted by NASA, is sophisticated enough to predict it with certainty. Doubly so when you consider that, despite, all the carbon dioxide, methane and chloroflourocarbons we've pumped into the atmosphere since the Industrial Revolution, U.S. and northern-latitude temperatures have actually declined over the last half-century.

The Greenhouse

The menace in question is called the "greenhouse effect," in which manmade pollutants rise into the atmosphere and prevent heat from escaping to space. NASA's Washington, D.C., forecast for 2050, for example, calls for a 1,000 percent increase in 100-degree days. The forecast gives a 70 percent chance that the polar ice cap will melt and swamp the Ocean City, Md., condos.

There is hardly a sane scientist who doubts that an increase in carbon dioxide will warm up an idealized, simplified Earth—the kind that exists in the computer's memory. The more CO_2, the warmer it gets. And there's no doubt that the amount of CO_2 in the atmosphere is going up.

Compounding the effect of CO_2 is the fact that a thin ice cap sits precariously atop our hemisphere, reflecting much of the sun's radiation. Melt it, and you have an ocean that absorbs a lot more radiation than it used to. Thus the Earth's surface warms, the Greenland ice sheets begin to melt, and the bottom falls out of the condo market in Ocean City.

From reading the newspaper accounts, one gets the impression that disaster has already begun. But buried in most expert testimony concerning the issue are more than the usual number of caveats. As NASA's Robert Watson told the Senate, "It's only a question of magnitude and time."

Little Evidence

Indeed. Here's the problem: In spite of the current increase in CO_2, and despite the headlines, there's precious little evidence that the Northern Hemisphere has warmed up significantly over the last 50 years. Moreover, when we look to the medium-high latitudes of our hemisphere, generally conceded to be most prone to climatic change, the warming simply isn't there.

A look at the temperature curves from the National Research Council's 1983 report "Changing Climate" sums up the problem. The northern latitudes, from 24 to 90 degrees north, have actually cooled off since 1930. The tropics warmed up slowly from 1880 to 1920, and then stopped. The Southern Hemisphere has warmed steadily throughout the period.

There is not one graphic in the entire 496-page report that indicates a statistically significant warming of the northern latitudes of our hemisphere since 1930. One figure shows two warm years at the end of the record, but to call that a significant indicator of the great warmup would be exactly akin to saying that the three consecutive very cold years in the late 1970s indicated imminent glaciation—which some people thought.

As far as the United States is concerned, the state-of-the-art temperature history remains Henry Diaz' and Robert Quale's 1981 article in the journal Monthly Weather Review. Although the last data in that article are from 1979, it is clear that mean U.S. temperatures have dropped in the last 30 years.

The most recent major article to detail the Northern Hemisphere's temperature history appeared in the February 1986 edition of the Journal of Climate and Applied Meteorology. Even though the data from the last few years are slightly biased toward warmer urban locations, a plot of the raw data does not support the conclusion that the big warmup has begun. While two very warm years highlight the end of the 135-year record, I challenge readers to find a statistician—even in Washington—who will say these indicate significant change.

Read one of the conclusions of the National Research Council in 1983: "In view of the relatively large and inadequately explained fluctuations over the last century, we do not believe that the overall pattern of variations in hemispheric or global mean temperature or associated changes in other climatic variables yet confirms the occurrence of temperature changes attributable to increasing atmospheric CO_2 concentration."

The conclusions of the 1985 self-proclaimed "state-of-the-art" report by the Energy Department aren't much different, although they are couched in the peculiarly Washingtonian neither-confirm-nor-deny mode:

"The findings from each of the lines of inquiry taken individually are, by themselves, insufficient to constitute convincing evidence that the climate models are correctly projecting the effects of the increasing CO_2 concentration on climate. However, to varying degrees, the evidence is generally consistent with, or at least not contradictory to, model projections of such effect."

The Northern Hemisphere

The real bad actors in the warmup scenario have been the Northern Hemisphere oceans. They simply refuse to go along with the program, showing nothing but a slow, significant cooling since World War II. With oceanic (as opposed to land) temperatures, short-term fluctuations are damped, although long-term changes take longer to appear.

So why isn't the Northern Hemisphere warming, if the amount of CO_2 in its atmosphere is increasing? The best explanation is that other factors, unknown or currently unexplainable, are interfering.

Example: Francis Bretherton (recognized by the atmospheric science community with numerous research awards and prizes)

42

and James Coakley of the National Center for Atmospheric Research recently looked again at satellite images to see if there has been an increase in Northern Hemisphere high cloudiness. According to John Botzum's Washington-based Weather and Climate Report, "Coakley said that an increase of 4 percent to 7 percent in cover by certain types of clouds could offset a doubling of carbon dioxide."

Cirrus clouds aren't the only unknown. Most of the computer models contain major limitations in their geography and oceanic heat transfer, and no global-climate model has accurately simulated changes in regional rainfall since the beginning of the increase in CO_2.

Meanwhile, the Southern Hemisphere has been warming up in a fashion that no one doubts. So what's the difference between hemispheres?

Perhaps some type of atmospheric turbidity—volcanic or manmade—has so far countered the effect of carbon dioxide in our hemisphere. The fact is that CO_2 stays in the atmosphere a lot longer than the stuff that goes into the air alongside it. While both are primarily produced in the Northern Hemisphere, only CO_2 resides long enough to diffuse into the Southern Hemisphere. Thus the Southern Hemisphere's clearer atmosphere receives and traps more radiation.

Sooner or later the CO_2 effect should predominate in our hemisphere, but probably not soon enough to turn Washington into beachfront property for a while. As NASA's Watson said, the magnitude and the timing are unclear.

Further, if such a thing can be said to be reassuring, Cassandra's Law of predicted atmospheric disasters says the first forecasts are usually the worst. Remember how we worried at first about the SST in 1970, and chloroflourocarbons in 1972 and the coming ice age in 1974? Even the nuclear-winter scenario has softened up a bit during the last year.

So until we understand why the Northern Hemisphere has been so slow to warm, please pass the research funding—and don't sell the condo.

GLOBAL WARMING: IDEAS IN CONFLICT

NO QUESTION ABOUT GLOBAL WARMING

Theodore K. Rabb

Theodore K. Rabb is a professor in the Department of History, Princeton University, Princeton, New Jersey. He is co-editor of Climate and History *(Princeton University Press, 1981).*

Points to Consider

1. Why is there no question about global warming?
2. Why has no action been taken?
3. What is the role of the scientist?
4. How can global warming devastate a social system?
5. What can be done to avert social and cultural disaster?

Excerpted from testimony by Theodore K. Rabb before the Senate Subcommittee on Environmental Pollution, June 11, 1986.

Trace gases are building up in the atmosphere at a geometric rate, and an unprecedented warming of the atmosphere may already be well under way.

A massive survey of research on climatic impacts, published last year, leaves its reader with disturbing news about the buildup of Carbon Dioxide in the atmosphere. There is no question that it is happening, that it is going to cause global warming, and that such changes in the past—even when they are of much lesser magnitude—have had widespread ill effects on human society. . . .

Unfortunately, a form of doomsday is what climate might have in store for us. Second only to nuclear disaster, changes in the environment are the most potent threat to the continued existence of society as we know it. Yet if the scientific community has a dismal record of rousing public understanding of the full, terrifying implications of nuclear arsenals—survivability is a word bandied about as it were a tourniquet for a minor wound—then on climate it has virtually no record at all. Despite studies by the bushel, and occasional—albeit usually highly technical—general discussions of what the future holds, there has been no visible effort to issue the clarion calls that might focus wide attention on the looming forces of climatic change. Average citizens may have some conception of nuclear war, but most would probably define the build-up of carbon dioxide, if given multiple choice, as some type of tooth decay.

This is where objectivity and the commitment to research for its own sake have brought us. The studies multiply. The fascinating problems are uncovered and dissected. Techniques of dazzling ingenuity are invented so as to derive—to give just one example—temperature records of more than a thousand years ago from the calcium deposits of water seepage in limestone caves. . . .

Action and Knowledge

The one inescapable result of all the work, regardless of specialty, is the discovery that trace gases are building up in the atmosphere at a geometric rate, and that an unprecedented warming of the atmosphere may already be well under way. Has that resulted in calls for public education, massive campaigns for remedial action, and fevered portrayals of the dislocations

Deja Vu?

By David Seavey, USA TODAY

or perhaps catastrophes that may ensue? Not at all. The few who *have* lit some beacons—Stephen Schneider's *Genesis Stategy,* with its predictions of ice caps melting and coastal cities inundated, is a rare example—are dismissed as insufficiently "scientific" and untrustworthy at best. Much preferred is a tame call for further study. If nobody can be 100% sure, then obviously the best tack—certainly for the scientists who will get the grants to do the research, as with "Star Wars"—is to keep at it until we *can* be 100% sure. Too bad if by then it's too late. At least the great scientific tradition of neutrality and restraint will have been preserved. Just one quotation will give the flavor of this outlook:

An increase in the average temperature by 3 or 4° C could lead to the beginning of an irreversible melting of glaciers. What will the properties of the new state of equilibrium of the biosphere be like; will they permit the existence of man? We do not know.

Know 100%? Perhaps not, but we have a pretty good idea. America's corn belt will no longer grow corn. It may grow in Saskatchewan, but there isn't much soil up there. Trees at home in the temperate zone will not flourish in their current habitat, but will someone have planted them all further north, in their new home? And will we be ready to move them again a few years later? That prospect, moreover, is not a comfortable two centuries away, as we once thought, but maybe just fifty years, and closing fast. Of course we do not know exactly what will happen, and when. But is that a reason for delay, especially since we already know a great deal—more than enough—and none of it is pleasant?

Evasions

The calm about all of this is shattering. And the evasions are extraordinary. In a world of fond hopes, perhaps the trend will not turn out as badly as the indicators now suggest, or the model will prove to have been too pessimistic. Maybe the effects will be cushioned by adaptations similar to those mankind has already shown itself capable of (hopefully without the enormous suffering those experiences brought in their wake). Could it be that we will all somehow muddle through, that it is really somebody else's problem—the politicians'?—or that opposing trends will cancel each other out? The latter, couched in such scientificese as "negative feedbacks" which "counteract any wide departure" or even "major excursions" from the norm, presumably require nature somehow to restore the equilibrium. How this might happen is not apparent, though there are those who believe that in fact we are on the brink of a major cooling, as the result of a cyclical return of the ice age, and maybe this would balance out the greenhouse effect. . . .

Scientific tradition, when applied to a subject as charged as this one, becomes an obstacle to understanding and action. The peril of general ignorance is beginning to outweigh the perils of alarmism.

Social Devastation

It is true that we also cannot predict exactly what the Carbon Dioxide buildup will change. But history and our own times are littered with societies devastated, destroyed, or merely damaged by climatic forces. From Sri Lanka around 1400 (when a stable society's agriculture was shattered by dwindling rainfall, it succumbed to previously weaker invaders, and cultural divisions were created that plague the island to this day) to Greenland around 1700 (when a flourishing European colony was decimated by declining temperatures), we can see entire populations either wiped out or forced by immense disruptions to move huge distances, totally reconstitute their economies and politics, and dismantle ways of life centuries old. There are dozens such examples, and they do not diminish in our own times, as mere mention of the words "Dust Bowl" or "Sahel" remind us. . . .

We look back at these eruptions. We see fumbling and usually futile efforts to come to terms with the upheavals, which in every case were totally unexpected. We see debilitating consequences that lasted generations. And yet none of these examples involves a climatic break from the past whose magnitude in the short term even approaches what most studies now tell us—not without warning, but with abundant warning—is likely to happen in the twenty-first century. Can we learn nothing from the past?

49

Role of Scientists

The ideal scenario would be for all the scientists who now have unmistakable evidence of accelerating Carbon Dioxide build-up and its future course to join together with the scientists who can determine how each 1° C of annual warming will affect the oceans, the land, and the air, and then unanimously tell the world, in the most dramatic tones possible, of the catastrophic dislocations humanity faces. But that is not what scientists do. Indeed, any attempt to organize such an effort would be viewed with suspicion. One alternative would be for an unimpeachable figure, revered as a leader beyond the range of the snipers, to make the move on behalf of his colleagues. The only time that ever happened was when Einstein wrote his famous letter of 1939 to Roosevelt about German nuclear research. Significantly, though, his later warnings about the dangers of the atom bomb went unheeded.

Yet the reason the Einstein letter worked was the identity of its recipient, and that is why this hearing can still be a beginning, regardless of the mores of the scientific community. What is needed above all is political leadership. In a situation lacking the obvious signs of disaster—nuclear reactors blowing up, tropical heat waves in February—but haunted instead by a distant and unseen menace, how else is the world to be galvanized? After all, it is no longer a scientific problem we are facing. The facts are basically in; we know in general terms what lies in store. The baton must pass to those who can make the issues salient and can convince society to face up to their demands—in other words, to our political leaders.

Conclusion

The tools they need are certainly at hand. Historians, sociologists, and anthropologists have studied dozens of peoples whose lives have been shattered by natural disasters both sudden and slow moving. They have analyzed strategies that have saved communities, and reactions that have merely made bad times worse. It is not too difficult, for instance, to learn why the potato fungus that caused starvation in Ireland in the 1840's had a far less malignant effect, in those same years, on the Dutch, who ate just as many potatoes as the Irish. And we have one enormous advantage even over the environmentally astute Dutch.

We have foreknowledge. We can therefore consider now, while there is still time, how we will address the issues that confront us. What *will* we want to do when Washington gets Miami's climate? Are such plans possible? Or do we want instead to seek out preventatives? Is massive, world-wide reforestation feasible? The only way we can answer questions like these is if, first, we admit they are of vital importance, and then, second, we get down to the task of finding out *how* to answer them. We may not know precisely where the quest will end, but it is urgent that we at least repeat the words first spoken not many yards from here; let us begin.

GLOBAL WARMING:
IDEAS IN CONFLICT

IMMEDIATE POLICY CHANGE
NOT NECESSARY

Thomas F. Malone

Dr. Thomas F. Malone made the following comments in his capacity as chairman of the Board of Atmospheric Sciences and Climate, National Research Council, National Academy of Sciences. Dr. Malone summarizes the conclusions of the Academy's 1984 report titled Changing Climate.

Points to Consider

1. Why is there no need for panic?
2. What might be the results of global warming?
3. What kind of energy policy should be followed?
4. What are the four major conclusions of the 500 page report on changing climate by the National Academy of Sciences?
5. Why is international cooperation important?

Excerpted from testimony by Dr. Thomas F. Malone before the House Subcommittee on Investigations and Oversight, February, 1984.

The evidence at hand does not support steps to change current fuel use from fossil fuels at present.

Our report is conservative in its conclusions, and will, we hope, abate some extreme negative speculations. In brief, we estimate that carbon dioxide will most likely double over the next century. This doubling will result in an increase in average earth temperature between 2 and 8 degrees Fahrenheit, with the lower range most likely. The temperature increase will, in turn, affect sea level, growing seasons, local water supplies, and climate patterns.

Despite the potential seriousness of some of these effects, our committee found the situation to be one of caution, not panic. We recommend expanding monitoring and continued research, but no immediate change in energy policy. One reason for this recommendation is as follows: There are many things we do not understand about CO_2 effects. There are other uncertainties about our future use of fossil and synthetic fuels.

For a Conservative View

In the committee's own words—"In our judgment, the knowledge we can gain in coming years should be more beneficial than a lack of action will be damaging. A program of action without a program of learning could be costly and ineffective." I hope that is impressed deeply. Watchwords for the immediate future should, in our committee's view, be research, monitoring, vigilance, and an open mind.

Adverse Effects

Now, some of the details of our findings. Among the adverse results that have been discussed is a major rise in sea level, about 2 feet over the next 100 years, due to melting of glaciers and expansion of sea water. This rate may increase in following centuries. This is clearly a serious prospect for low-lying areas of the world like Florida, Holland, Bangladesh, but defensive measures seem feasible. A 20-foot rise due to breakup of the West Antarctic ice sheet would take several hundred years, after its surrounding ice shelves had receded. Now to place these changes in context, you might recall that the sea level has risen only about 6 inches in the last century but 500 feet since the last glacial period about 15,000 years ago.

A second potential adverse effect is on agriculture. While predictions of global warming are probably quite reliable, predictions of specific regional climate changes are much less certain. Nevertheless, regional changes will occur and may have serious impacts. Reports of estimates of the aggregate effect on U.S. agriculture through the end of this century indicate that the negative impact of changing climate will be largely balanced by the positive effect of increased fertilization due to increased CO_2. With the demonstrated ability of the U.S. agricultural complex to adapt to changing conditions, yields can be maintained or increased and we predict no overall threat to American agriculture over the next few decades.

I would stress that the real central issue here is the rate at which the thing called technology per year can increase productivity compared to the rate of change, so we are not taking a big jump of 2 to 8 degrees. It's the yearly change that is the critical factor, and that is often overlooked in impact studies. The most serious effect would be in the arid regions, and even in our own West a slight warming and a decrease of rainfall would, if it occurs, slow stream runoff and could have severe effects.

Four Conclusions

Now that is the thrust of our report, all 500 pages, but I would mention four of what I feel are principal conclusions. The first is that priority attention should be given to long-term options that are not based on combustion of fossil fuels. To be specific, I feel that we should pick up the second generation of the pathbreaking study by Wolfe Haefule at the International Institute of Applied Systems Analysis, where he and his international group thought deeply about how we get from here to there, and that's the kind of thing we should be doing.

Secondly, the evidence at hand does not support steps to change current fuel use from fossil fuels at present; and, third, it is possible—possible—that steps to control costly climatic changes should start with the non-CO_2 greenhouse gases. We need to understand those better. They are looming as increasingly

Time to Adjust

Atmospheric buildup of carbon dioxide could cause serious climate changes in the next century, but there is no basis for panic or a precipitous shift in energy policy, the National Academy of Sciences said Thursday.

The 500-page report on the greenhouse effect takes a less pessimistic view than another released earlier in the week by the Environmental Protection Agency.

The EPA's estimates of rising oceans and warmer climates generally jibed with those of the academy, but the agency warned the consequences could be "catastrophic" unless the world begins planning for the shifts "with a sense of urgency."

Associated Press, October, 1983

important, and they may be more amenable to control than carbon dioxide.

The fourth conclusion is that this is intrinsically an international problem and requires an international network of science, scientists, convergent with these problems. I stress the importance of achieving a consensus, if possible, within the international scientific community before governments start negotiating how they would cope with this problem. For example, if we were able to develop a photovoltaic method of energy, it could conceivably reduce our demand for coal, which would decrease the price of coal, which would increase the consumption of coal in other countries, so that one country cannot address this problem by itself.

We have in place, and I will be meeting in 2 weeks in Hangzhou, China with a group of scientists, including Chinese, Russian, English, Swedish, to discuss the strategy of a world climate research program. This is a joint enterprise of the International Council of Scientific Unions, the World Meteorological Organization, and the United Nations Environmental Program, and one of the topics is this question of CO_2-induced climatic change, and this is a healthy trend. I met in August with a group in Villach, Austria, looking at the impact. Again, we had representation from these other countries. I will come back to the international in my recommendations.

Monitoring

A few recommendations: We will have a symposium in Ottawa in September, bringing together about two dozen scientists from all over the world to discuss a monitoring program which would link together the ocean, the atmosphere, the biosphere, forests, the land, and the solar-terrestrial domain. This is the kind of program that I think would be responsive to the needs expressed earlier this morning.

It is clear to me that the importance of developing the potential of the earth-looking satellite, which has lagged in comparison to astronomical purposes and planetary science purposes, will be a very powerful tool for examining the ocean and for examining the land and the forest. There are major programs: The World Ocean Circulation Experiment, called [WOCE], W-O-C-E; the Tropical Ocean Global Atmosphere with the acronym [TOGA], T-O-G-A, addresses specifically the type of El Nino thing that has been referred to. These are programs just getting underway and they deserve our support.

A second emphasis should be on the non-CO_2 gases. With respect to impact studies, it's clear that the sea level problem and agricultural deserve high priority attention. With respect to emission studies, the kind of models that Dr. Nordhaus developed—and I particularly commend to your attention that chapter in the National Academy of Sciences report written by Dr. Nordhaus, which is a pioneering effort to develop a sound economic emission type of model, and it is also, I believe, supplemented and supported by the work of Edmondson and Reilly at Oak Ridge—these are efforts that should be expanded. We need a larger community, rather than the three or four people that are addressing this now.

I would also commend to you the chapter in that report by Professor Tom Schelling of Harvard, which analyzes very thoughtfully some of the policy implications.

Finally, I would say two things which are, in part, personal observations from what I have heard this morning. I think that DOE (Department of Energy) might well institute the kind of peer review program that the National Science Foundation uses

Illustration by Craig MacIntosh

Reprinted with permission from the *Minneapolis Star and Tribune*

to ensure the highest quality in the research it supports, and
in conclusion I would plead for a balance among the topics
emissions, carbon cycle, climatic change, environment impact,
social, economic, and policy considerations.

GLOBAL WARMING: IDEAS IN CONFLICT

COST OF INACTION WILL BE ENORMOUS

Michael Oppenheimer

Dr. Michael Oppenhiemer is a senior scientist with the Environmental Defense Fund. His expertise is in the area of atmospheric physics and chemistry. His recent research has focused on pollution and acid rain.

Points to Consider

1. How is the "greenhouse effect" explained?
2. Why will global warming be a great disaster?
3. Could there be any positive results from global warming?
4. What immediate actions should governments take?
5. What should the United States do?

Excerpted from testimony by Michael Oppenhiemer before the Senate Committee on Environment and Public Works, June 11, 1986.

Our knowledge is sufficient to demonstrate large potential risk. In such circumstances, prudence argues for four immediate actions.

My comments focus on climate change, a problem which, if unchecked, will come to dominate all others in its effect on the environment. The composition of our atmosphere, the earth's radiation balance and our climate, are changing due to human activity. The changes humans are bringing to the atmosphere will shortly begin to react on the biosphere. From the perspective of human history, these changes will be rapid and costly; and many will be highly undesirable. The viability of many ecosystems is at stake as is, some say, the viability of civilization as we know it. The changes may occur faster than our knowledge of them increases; yet, we currently know well how to limit these changes. Since the consequences of ignoring climate change will be severe, it is time for the U.S. government, along with governments of other nations, to come to grips with this problem.

We have begun an unintended experiment on the atmosphere and, eventually, the biosphere. As with all experiments, we know some of the questions to ask, but we will have limited insights into the answers until the results are in. Unfortunately, large consequences will have become inevitable by that time. On the other hand, this experiment can be altered while in progress to limit those consequences, even though this course requires action with only partial understanding. We cannot afford an undesirable outcome; we cannot afford to leave the experiment unchecked. It is time to develop policies to limit climate change. . . .

Changing Climate

The climate of the earth is determined by the balance between the rate at which the earth is heated by solar energy and the rate at which it is cooled by radiating heat into space. This balance is strongly affected by various components of the atmosphere which can act as a blanket and slow the outflow of radiating heat (by absorbing infrared or heat radiation). Such a slowdown can warm the planet. Among the natural atmospheric constituents capable of altering the radiation balance are water vapor, carbon dioxide and methane. Artificial substances, such

59

as fluorocarbon compounds (which can also alter the earth's ultraviolet shield of ozone), can act similarly, even in minute quantities.

Recent atmospheric measurements reveal that the abundance of several of these chemicals, particularly carbon dioxide and methane, are increasing. The carbon dioxide increase is attributable to fossil fuel combustion, and to the destruction of forests, which converts carbon from organic forms into carbon dioxide. Fossil fuel "mining" and combustion is also responsible for some of the increases in methane, nitrous oxide and other "greenhouse" gases. Some of the increases are due to other human activity (such as refrigeration, which releases fluorocarbons and agriculture, which releases some methane). Emissions of carbon dioxide and several other greenhouse gases are growing.

The effect of these human activities on climate can be predicted with computer models. The picture painted by these studies is for an earth, within the next century, which is climatologically very different from the one we know. . . .

Effects of Climate Change

Shifts in precipitation patterns can be expected to bring arid conditions to the mid latitude breadbasket areas while increasing precipitation in continental areas at more northerly latitudes. Sea levels will rise slowly at first, and more rapidly after the next century if major ice shelves destabilize and slip into the sea. Coastal flooding could become significant in already marginally useable terrain in places such as Bangladesh and coastal Louisiana, as early as the mid 21st century. The frequency of temperature extremes may increase much faster than the mean so that many more very hot days can be expected in Mid America within a few decades.

If fossil fuel use continues only at current rates for the next 100 years, marked climate changes will occur. However, if fossil fuel use and other activities which produce trace gases continue to increase at current rates of change, the climate will be greatly altered in far less than 100 years.

In either case, if the changes in temperature and precipitation predicted by the atmospheric models occur, the effects on the global ecosystem will be substantial. Some coastal wetlands will disappear and there is no certainty of their re-creation further inland. Coastal habitation and other infrastructure will be destroyed in low-lying areas. In areas such as Bangladesh where the

By Bill Sanders, *Milwaukee Journal*
Permission of News America Syndicate

problem of limited arable land was made apparent by the recent
cyclone, the population will be further compressed. In the developed world, agricultural productivity will decline in some areas
of current high productivity and moderate rainfall, as they become
arid. Whether other areas will become wet and fertile rapidly
enough to avoid major dislocation is problematic. Entire ecosystems, such as those of the Arctic, may be eliminated.

On the other hand, some benefits may accrue from climate
change. For instance, the moderation of the northern climate
could allow increased habitation and cultivation in parts of Canada
and the Soviet Union. Some crops may benefit from high temperature and carbon dioxide levels. No quantitative comparison has
been made of the practicality or costs of preventive vs. adaptive
strategies. Nor is such a comparison entirely feasible, as we
cannot properly value global scale ecosystem loss.

A small scale preview of potential dislocating effects of climate
change was evident in one short month last spring in three
climate induced occurrences unrelated to greenhouse warming.

Previously lush forests burned to the ground due to a sustained drought in Florida. The Northeastern U.S. suffered a sustained drought threatening urban populations with at least minor economic dislocations due to water shortage; large numbers of people died in a cyclone in Bangladesh which brought up the sea level a few feet. While sea level rise due to the greenhouse effect would be more gradual, the loss of agricultural land may be even more extensive and permanent.

Several important points should be made: the predicted warming would produce global mean temperatures warmer than any during human experience and the change in temperature exceeds temperature variations during recorded history. We will transcend human experience long before a CO_2 doubling occurs. Furthermore, weather events, such as successive high temperature days, will deviate from the past by larger amounts than will the mean. . . .

Immediate Action

Our knowledge is sufficient to demonstrate large potential risk, but our level of ignorance is very high. In such circumstances, prudence argues for four immediate actions:
—Slow the growth in emissions of greenhouse gases to slow climate change, through vigorous attention to end use efficiency in energy and materials use and through other measures;
—Act to preserve forests and support reforestation to protect the terrestrial carbon reservoir;
—Increase massively the support for research on ecological impacts of all of these insults, to allow our knowledge to change faster than the climate;
—Slow the alteration of forest ecosystems by reducing air pollution stresses.

What Government Can Do

The remoteness and scope of climate change has discouraged policy makers from coming to terms with its potential effects. The alternative courses available, prevention, adaptation, or some mixture of these approaches, all require the focused attention of governments, and international cooperation. With some climate change probably already occurring, remoteness is now an insufficient reason for inaction.

Changing Rapidly

With unusual unanimity, scientists testified at a recent Senate hearing that using the atmosphere as a garbage dump is about to catch up with us on a global scale. The Earth's climate is changing rapidly and its protective shield of ozone is shrinking.

Michael Oppenheimer, *New York Times,* July, 1986

A fundamental characteristic of climate change is that the greenhouse gases are relatively non-local in nature. Due to atmospheric mixing, location of source is not related to climate effect for CO_2 and most other greenhouse gases. Thus, climate change limitation can only come through international cooperation on greenhouse gas limitation. However, the industrial nations of Europe and the United States have a unique lead role in such a cooperative venture, since they are the current source of the bulk of emissions. If those nations do not take a leadership role now, substantial climate change will become inevitable as Third World countries develop, and increase their greenhouse gas emissions.

Preventive Strategies

Preventive strategies in particular, only can be accomplished internationally and within a context which will allow economic growth in underdeveloped nations. Adaptive strategies, which could involve migration and changes in food distribution and in infrastructure, also require international cooperative efforts. A conservative view of global ecosystems favors preventive strategies, but with some climate change inevitable (and perhaps already occurring), strategies which are broadly protective, focusing on climate change minimization but preparing for some adaptation, merit attention.

A successful approach to climate change policy should aim for an international accord on greenhouse gases. Such an accord will only develop when the scientific community presents a consistent, understandable scientific position which elaborates the scope of impending climate changes and their global ecological consequences. The development of such an accord provides the occasion to balance the virtues of the various strategies and determine a realistic degree of prevention through greenhouse gas limitation. . . .

Private Organizations

Various private sector organizations, including the Environmental Defense Fund and the Beijer Institute of the Royal Swedish Academy of Sciences, are working to move the international community to the determination of policy. A meeting of scientists and policy makers is planned for the summer of 1987 to accelerate this process. These efforts are supported in part by private American foundations.

The Environmental Defense Fund supports the development of a timetable for activities leading to a convention on greenhouse gases. These activities should include specific policy research projects, deliberations among scientists and policy makers at the international level, and the scheduling of activities to develop a convention by a specific date.

The United States

With activity already underway, the U.S. government should immediately:
—support international efforts to develop a greenhouse gas convention;
—develop policy alternatives for limiting greenhouse gases and protecting forests. Such policy development should include estimates of the costs of inaction;
—massively increase support for climate change research in general and ecological research in particular;
—push for meaningful limitations on fluorocarbon production in the context of the 1987 Convention;
—encourage actions internally which increase end use efficiency in energy and materials use;
—attend to the long overdue alleviation of other regional air pollution stresses, such as acid deposition and regional smog.

With regard to climate change, we cannot afford to just "let it happen." The costs of a non-policy will be enormous. Let us set out now to determine a reasonable course for greenhouse gas limitation before we are overtaken by the dire consequences of inaction. Otherwise, unacceptable levels of climate change may be "in the bank" before we even understand what we have wrought.

GLOBAL WARMING: IDEAS IN CONFLICT

WE LACK THE ABILITY TO PREDICT CONSEQUENCES

Wallace S. Broecker

Wallace S. Broecker is a member of the National Academy of Sciences and a professor of geochemistry at Columbia University.

Points to Consider

1. Why do we lack the ability to predict the consequences of the greenhouse effect?
2. Why could there be large variations in weather patterns?
3. What climatic role is played by the oceans?
4. Why has our view of world climate been too simplistic?
5. How can better management of federal agencies with research money promote progress?

Excerpted from testimony by Wallace S. Broecker before the Senate Subcommittee on Environmental Pollution, June 10-11, 1986.

I maintain that the information in hand is inadequate and therefore attempts to squeeze valid predictions from it are destined to failure.

In my estimation we currently lack the ability to reliably predict the consequences of the buildup of greenhouse gases in our atmosphere. Furthermore, at the current rate of progress, this deficiency will remain well into the twenty-first century. Hence, many of the important consequences of the greenhouse buildup will be upon us before we are even partially prepared to cope with them. The reason this situation exists can be stated very simply. The problem is a tough one. It will certainly not be solved in a few years or even in a few decades. In fact, it may never be fully understood. Like cancer and military defense, a highly sophisticated long-term effort will have to be made if substantial progress is to be achieved. While society has accepted the reality of the situation regarding cancer and military defense, it has not come to grips with either the importance or complexity of the situation relating to future global climates. Certainly the Department of Energy which leads our nation's CO_2 program has yet to be enlightened. Its program is run as if it were a routine engineering problem fully soluble on a ten-year time scale.

Inadequate Information

The CO_2, CH_4, freons, etc. that we are putting into the atmosphere will surely bring important changes in the environmental geography of our planet. While we are not likely to be able to prevent these changes, we can certainly better prepare ourselves to cope with them. However, if we are to succeed in developing response strategies, our attitudes and approach must be radically changed. The main problem with the present approach is that the underlying assumption is that we have most of the information we need and that the answers we seek are obtainable by massaging this information and making computer simulations capable of matching it. I maintain that the information in hand is inadequate and therefore attempts to squeeze valid predictions from it are destined to failure.

While the research during the last decade has not pushed us ahead very far in our ability to predict the consequences of the greenhouse buildup, it has dramatically shown us that the climatic system is interconnected in ways we have not seriously considered and that these interconnections lead to the possibility of responses about which we have not even dreamed. If this evidence is correct, it tells us that the "general circulation models" which are now used to predict the climatic impacts of a greenhouse buildup give us a very conservative view of how climate will change. The new evidence opens the possibility that rather than paralleling the smooth buildup of greenhouse gases, future climate changes may come in sharp steps and these steps may have complicated geographic patterns. If so, then we must begin to think in terms of climate surprises that may, without warning, cause shifts in temperature and rainfall pattern.

What I have just said is not found in most reports on the possible consequences of the greenhouse buildup. One reason is that these ideas are new. Another is that these ideas constitute a rather venturesome extrapolation of the evidence in hand. I am confident enough in them, however, to bring them to your attention as a serious possibility.

An Irregular Climate

What is the evidence which leads me to these conclusions? It comes from the paleoclimatic record of the last 100,000 years. In particular, it comes from borings made through the Greenland ice cap. The ice record shows unequivocal evidence for sudden large jumps in climate. Analysis of this ice core evidence along with that from deep sea cores raised from the floor of the northern Atlantic Ocean and from cores obtained from European bogs, a consistent picture emerges which points toward the causal factor. This causal factor turns out to be changes in the mode of operation of the ocean's large scale circulation system. Thus, we have an indication of the possibility that when the climate system is provoked, the large-scale features of the Earth's great ocean-atmosphere heat engine may reorganize into a pattern more suitable to the new conditions. As these reorganizations alter the pattern and magnitude of heat transport within the ocean, they alter the climate on the adjacent continents.

Another interconnection about which we have recently become aware involves atmospheric chemistry and ocean biology. Associated with the sharp climate changes recorded in Greenland ice are jumps in the CO_2 content of the atmosphere. While the exact origin of these jumps is still the subject of debate, those working on the problem agree that it must involve major changes in the biogeochemical cycles occurring within the ocean and that these in turn stem from the reorganization in large scale ocean circulation pattern.

A Simplistic View

In my estimation, these new findings warn us that our view of the situation has been too simplistic. We have relied on simulations which, by their nature, prevent what may be the most bothersome aspects of the future climate response from occurring.

What should we do? My feeling is that we should launch efforts to understand those elements of the climate system which are now poorly understood. In order of importance, I would list the elements as follows:

1. The large scale circulation of the ocean.
2. The processes regulating soil moisture.
3. The processes responsible for cloud formation.
4. The role of global biogeochemical cycles in determining the trace gas content.
5. The processes regulating sea ice extent.

Of course, programs already exist in all these areas but in my estimation they will not bring the desired answers on the appropriate time scale. In each area we need major new observational programs to supply key data needed to develop a better physical understanding. In each area we need a cadre of young scientists with the appropriate training.

We also need to intensify our study of climate changes which have taken place over the last 100,000 years. Nature on her own has conducted climate experiments of large magnitude. The response of the system to these experiments is recorded in sediments and in ice. By thoughtful study of these records, we may be able to reveal modes of interaction among the various major elements of the climate system that would not otherwise come to mind. These hidden interactions are likely to carry the greatest threats.

Research Money

Progress on this as well as other long term environmental questions (ozone, acid rain, water quality, etc.) is greatly impeded by the manner in which authority is distributed among federal agencies. Much of the money for research now lies in the hands of the Departments of Agriculture, Energy, and Environment. These agencies find it difficult to fund the long term programs which are required if we are to push forward our understanding of the systems we wish to protect. Instead, they fritter away large amounts of money on short term projects which do little to remove our basic ignorance. Furthermore, jealousies among the agencies greatly hinder attempts to put into action those long term programs which individual agencies deem worthy.

I see only one workable solution to this festering problem. Congress should create an entity charged with conducting the long term research projects which are essential to the wise management of the environment. This entity must be isolated from the immense political pressures which buffet the agencies responsible for environmental regulation. It must also have a mechanism to generate cooperative ventures with the agencies controlling the domains to be studied.

Conclusion

The thrust of my testimony is that the potential consequences of the buildup of greenhouse gases (CO_2, CH_4, freons, etc.) in our atmosphere are not being given adequate attention. While the impacts of this increase remain uncertain, there is no doubt in my mind that they will be of great importance to agriculture and wild life. Mankind must prepare to cope with negative aspects of these changes and to take advantage of the positive ones. The complexity of this task is enormous rivaling other major challenges confronting mankind (i.e. the development of a long term energy supply, the control and cure of cancer. etc.).

As for these other quests, the route to success is by no means clear. Unfortunately, the management of the DOE CO_2

70

program treats the greenhouse problem as an engineering task rather than as a profound scientific challenge. They think mainly in terms of short-term goals (which they refer to as deliverables) instead of building a long-term ediface of people and observations which will lead to the breakthroughs we so desperately need if we are to properly prepare ourselves for the environmental changes which will come over the next hundred or so years.

If progress is to be made, the responsibility for the program must be placed into the hands of a qualified scientific manager who will listen to the knowledgeable people in the field.

GLOBAL WARMING: IDEAS IN CONFLICT

THE EVIDENCE IS IN

Rafe Pomerance

Rafe Pomerance is the president of Friends of the Earth, an international environmental organization with 30,000 members in the United States and affiliates in 22 nations.

Points to Consider

1. How are the reports on the greenhouse effect by the Environmental Protection Agency and the National Academy of Sciences explained?
2. Why is there hope for dealing with future problems of global warming?
3. What new technologies can help reduce environmental pollution?
4. How will the material well being of rich industrial nations be affected by a policy of conserving energy?
5. What policies has the U.S. followed?

Excerpted from testimony by Rafe Pomerance before the House Committee on Investigations and Oversight, February, 1984.

The burden of proof of safety falls on those who propose continuing present energy and agricultural policies until "all the evidence is in". By then it will be too late to cancel the experiment.

I think it is time to act. In fact, it's really too late to avoid initial warming. We know what to do. The evidence is in. The problem is as serious as exists. People talk about not leaving this to their grandchildren. I'm concerned about leaving this to my children. . . .

Scientific Reports

The Environmental Protection Agency (EPA), and the National Academy of Sciences (NAS) released independent studies on the "greenhouse effect"—the warming of the atmosphere caused by the release of carbon dioxide (CO_2) from burning coal, oil, and natural gas, as well as tropical deforestation, and industrial emissions of other trace "greenhouse" gases. EPA foresees a rise in atmospheric temperatures as early as the 1990's reaching major proportions early in the next century. The NAS puts the date for significant climate change in the mid-twenty-first century.

This atmospheric warming could easily bring us a climate averaging 4°C warmer than today but the warming would be three to five times greater in polar regions. These increasing average temperatures, together with reduced temperature differences between the equator and arctic regions, will bring large shifts in rainfall patterns. . . .

In an overcrowded, overarmed world, the disruption of food supplies, or the loss of crop and coastal lands could well lead to widespread wars including the strong possibility of accidental superpower conflict when climate change reinforces other pressures on natural resources. In the final analysis the latter risk is probably the greatest danger of world climate change. While it may be possible, though probably not desirable, to adapt to some climate change, history suggests that such transitions are often accompanied by conflict—a global disaster in the nuclear era if superpowers were involved would be highly probable.

When it comes to recommendations, the two reports draw opposite conclusions. The Environmental Protection Agency says it is too late to stop global warming and that the best policy is to adapt. The National Academy of Sciences concludes that we do not yet know what to do. It recommends waiting while further studies are conducted. The truth is neither. We must prepare for some climate change while also acting NOW to prevent drastic climate modification which will result if current energy, industrial and agricultural policies continue. . . .

A Certain Outcome

Surely no responsible scientist would conduct an experiment with such enormous consequences without being reasonably certain of the outcome. Yet, although the National Academy acknowledges great uncertainties in this "unintentional experiment", it fails to recommend cancelling the experiment until we have more information.

We *should* conclude just the opposite. . . .

The burden of proof of safety falls on those who propose continuing present energy and agricultural policies until "all the evidence is in". By then it will be too late to cancel the experiment. We must act now to reduce the use of fossil fuels until we know what the results will be, not *vice versa*. Of course, that policy would serve many other ends such as reducing acid rain, increasing energy security, and promoting the development of energy conservation, as well as solar and renewable energy for the future when oil is scarce.

A Policy of Hope

Contrary to EPA's initial assessment, there is hope. We can at least avoid a major greenhouse effect even if some climatic change is in the works. Catastrophic warming is neither an act of God, nor man's fate because determinate economic models tell us so. Rather, it is a direct result of energy, population, agricultural, and industrial policies. We need to implement energy and population policies that will reduce world fossil fuel consumption. And we must stop the pollution produced by manufacturing and industrial agriculture, which is the source of other trace greenhouse gases, such as freons, methane, nitrous oxide, and ozone. Some of these gases are already known to cause over $2 billion in crop losses annually and could themselves drive a substantial climate warming even if we drastically reduce fossil

energy use. Much more research must be funded to discover
the sources and potential danger of these trace gas emissions.

In its projections of future fuel needs, EPA assumes that an
extra dollar of income on industrial production will create a unit
of energy demand in a lockstep. For example, the study projects
that through the growth of per capita GNP, Americans will consume
325% as much food, clothing, housing, motor vehicles, and similar
products in 2050 as today. That projection is more a Dow-Jones
forecast of health spa stocks than an accurate measure of our
nation's girth! It overlooks the effect of consumer saturation (do
we really need three cars per person?), reduced working hours,
and structural shifts from a predominantly industrial to an informa-
tion-intensive economy.

EPA also assumes the Western style development of the Third
World. For example, Africa would have a per capita annual
income of $1940 in 2050 as opposed to $340 in 1975. Natural
resource limitations make such growth extremely unlikely. Clearly,
while economies will grow, increases in the standard of living
must come primarily in the quality, not quantity, of life—for many
reasons, of which climatic change, as well as the depletion of
soil and other natural resources are but partial constraints.

For the early twenty-first century, at which time we could now significantly affect energy supply and demand over the whole economy, there are a wealth of new money saving high efficiency technologies using a fraction of the energy we consume today. Many recent studies, including the Solar Energy Research Institute's (SERI) "Sawhill" study have found that we can substantially reduce energy consumption even while the economy keeps growing.

For example, in the U.S., Canada, and Europe, contractors are now building houses that consume only 5-10% of the average heating energy used today. In Canada there are "zero-energy" office buildings which live off the heat produced by their lighting, by heat produced by office equipment, and the body heat of their occupants. Car manufacturers have demonstrated 80-100 mile per gallon versions of existing subcompacts. At 80 mpg, during their lifetime, these cars would save 5,300 gallons of gasoline and approximately $6,600 against the typical car on the road today which averages 15.3 miles per gallon. That represents an 80% reduction on CO_2 emissions and a corresponding increase in energy efficiency. For all classes of vehicles, conversion to high energy efficiency would bring major savings to consumers.

Renewable Energy

Similar progress has been made in the development of renewable energy. Here the U.S. is in a leading position to open potential worldwide markets for its industry.

Together, high efficiency and renewable energy offer the opportunity to reduce fossil fuel consumption to a fraction of today's level. The possibilities have been set forth in detail by my colleagues, Amory and Hunter Lovins, Florentin Krause, and Wilfred Bach in their report commissioned by the West German government, and published in the U.S. as *Least Cost Energy,* as well as other publications of the *International Soft Energy Project*. These two strategies could delay further atmospheric changes enough to safely conduct a hundred years of research on the greenhouse effect while we develop the political and economic institutions to responsibly exercise climatic stewardship in the twenty-first century—sixteen years from now. In that case, climate

Fear of Frying

While most environmental problems do not necessarily have the potential to destroy civilizations, and appear to be controllable if political processes work fast enough, this threat is different. The Environmental Protection Agency says that the greenhouse effect—which occurs because a byproduct of combustion, carbon dioxide, allows the sun's rays to penetrate the earth's atmosphere but prevents infrared radiation from escaping it—is just beginning in earnest. And it can't be stopped, or even slowed, without the most drastic measures.

The Nation, May 26, 1984

change would be largely confined to the limited warming that will occur if all CO_2 emissions ceased tomorrow.

But the new generation of energy technologies will only be in place on time if public and financial policies mobilize the necessary research and capital so that consumers will use them. We need to motivate appliance, automobile, and equipment manufacturers to produce high efficiency products beginning now. Consumers will only pay the higher capital costs if they understand their economic interest in doing so and if it is easy to purchase, pay and service these technological improvements. One model is that of electrical utilities which promote and finance moneysaving energy conservation. But even California, which has been a leader, has a long way to go. For example, it would pay California electrical utilities to give a free high efficiency refrigerator to every family in the state while saving the need to construct 1700 megawatts of generating capacity. In the Midwest a similar strategy could help to solve the acid rain problem by reducing the need for stack scrubbers while also limiting CO_2 emissions.

Market Competition

For the most part, implementation of this climate-saving strategy can rely on market competition. Internationally, as Harvard Business School's energy experts, Robert Stobaugh and Daniel Yergin have pointed out, the United States exercises a strong market demand pull once the momentum for conservation and renewable energy is established. Such leadership is all the more important since developing nations' energy consumption is now strongly shaped by the products and technologies provided by industrial nations.

On the average over the next five decades, industrial nations could have the same standard of living while using about 20% of the energy per person that we use today. And we could save money. As noted earlier, investing in these improvements is far cheaper than building new power stations and synthetic fuel plants. We should support redirecting money appropriated for the Synfuels pork barrel to local, regional and Federal programs for energy conservation and renewable sources.

Unfortunately, the Reagan administration has dismantled or attacked every Federal energy program in the area of solar and conservation. Federal energy policies must be reversed. If we are incapable of doing so now when it would pay to do so and would provide greater long term energy security, can we hope to mobilize the discipline and sacrifice demanded without conflict in a world inundated by unprecedented climatic changes?

RECOGNIZING AUTHOR'S POINT OF VIEW

This activity may be used as an individualized study guide for students in libraries and resource centers or as a discussion catalyst in small group and classroom discussions.

The capacity to recognize an author's point of view is an essential reading skill. Many readers do not make clear distinctions between descriptive articles that relate factual information and articles that express a point of view. Think about the readings in chapter two. Are these readings essentially descriptive articles that relate factual information or articles that attempt to persuade through editorial commentary and analysis?

Guidelines

1. The following are possible descriptions of souces that appeared in chapter two. Choose one of the following source descriptions that best defines each source in chapter two.

Source Descriptions

a. Essentially an article that relates factual information
b. Essentially an article that expresses editorial points of view
c. Both of the above
d. Neither of the above

Sources in Chapter Two

——Source Six

"Facts Fail to Back Warming Theory" by Patrick J. Michaels.

——Source Seven

"No Question About Global Warming" by Theodore K. Rabb.

——Source Eight

"Immediate Policy Change Not Necessary" by Thomas F. Malone.

——Source Nine

"Cost of Inaction Will Be Enormous" by Michael Oppenheimer.

——Source Ten

"We Lack Ability to Predict Consequences" by Wallace S. Broecker.

——Source Eleven

"The Evidence Is In" by Rafe Pomerance.

2. Summarize the author's point of view in one to three sentences for each of the readings in chapter two.

3. After careful consideration, pick out one reading that you think is the most reliable source. Be prepared to explain the reasons for your choice in a general class discussion.

CHAPTER 3

THE OZONE LAYER

OVERVIEW

OZONE DEPLETION AND
THE GLOBAL CONSEQUENCES

Robert T. Watson

For several decades scientists have tried to understand the complex interplay among the chemical, radiative, and dynamical processes that govern the structure of the Earth's atmosphere. During the last decade or so there has been particular interest in studying the processes which control atmospheric ozone since it has been predicted that human activities might cause harmful effects to the environment by modifying the total column amount and vertical distribution of atmospheric ozone. Most of the ozone in the Earth's atmosphere resides in a region of the atmosphere known as the stratosphere. The stratosphere extends from about 8 km at the poles, and 17 km at the equator, to about 50 km above the Earth's surface. Ozone is the only gas in the atmosphere that prevents harmful solar ultraviolet radiation from reaching the surface of the Earth. The total amount of ozone in the atmosphere would, if compressed to the pressure at the Earth's surface, be a layer about one eighth of an inch thick. Figure 1 schematically illustrates the vertical distribution of ozone and temperature. Unlike some other more localized environmental issues, e.g. acid deposition, ozone layer modification is a global phenomenon which affects the well-being of every country in the world. Changes in the total column amount of atmospheric ozone would modify the amount of biologically harmful ultraviolet radiation penetrating to the Earth's surface with potential adverse effects

Excerpted from testimony before the Senate Committee on the Environment and Public Works, June, 1986.

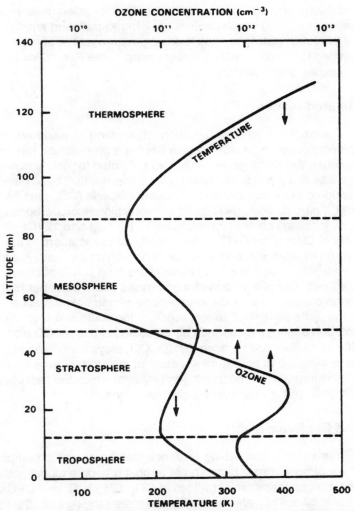

Figure 1. Temperature profile and ozone distribution in the atmosphere.

U.S. Senate Committee on Environment and Public Works

on human health (skin cancer and suppression of the immune response system) and on aquatic and terrestrial ecosystems. Changes in the vertical distribution of atmospheric ozone, along with changes in the atmospheric concentrations of other infrared-active (greenhouse) gases, could contribute to a change in climate on a regional and global scale by modifying the atmospheric temperature structure which could lead to changes in atmospheric circulation and precipitation patterns. The so-called greenhouse gases are gases which can absorb infrared radiation emitted by the Earth's surface, thus reducing the amount of energy emitted to space, resulting in a warming of the Earth's lower atmosphere and surface.

Related Issues

The ozone issue and the greenhouse warming issue are strongly coupled because ozone itself is a greenhouse gas, and because the same gases which are predicted to modify ozone are also predicted to produce a climate warming. These gases include carbon monoxide (CO), carbon dioxide (CO_2), methane (CH_4), nitrous oxide (N_2O), and several chlorofluorocarbons (CFC's), including chlorofluorocarbons 11 (CFC_{13}) and 12 (CF_2C_{12}). CH_4, N_2O, and the CFC's, respectively, are precursors to the hydrogen, nitrogen, and chlorine oxides which can catalyze the destruction of ozone in the stratosphere by a series of chemical reactions. Concentrations of these gases in the parts per billion range control the abundance of ozone whose concentration is in the parts per million range, e.g. one molecule of a chlorofluorocarbon destroys thousands of molecules of ozone. CO and CO_2 can affect ozone indirectly. . . . CO_2 plays a key role in controlling the temperature structure of the stratosphere which itself is important in controlling the rates at which the hydrogen, nitrogen, and chlorine oxides destroy ozone.

The Evidence

There is now compelling evidence that the chemical composition of the atmosphere is changing at a rapid rate on a global scale. The atmospheric concentrations of CO_2, CH_4, N_2O, and CFC's 11 and 12 are currently increasing at rates ranging from 0.2 to 5.0% per year. The concentrations of other gases important in the ozone and global warming issues are also increasing, some at an even faster rate. These changes in atmospheric composition reflect in part the metabolism of the biosphere and in part a broad range of human activities, including agricultural and combustion practices. It should be noted that the only known source of the CFC's is industrial production. They are used for a variety of uses, including aerosol propellants, refrigerants, foam blowing agents, and solvents. At present one of our greatest difficulties in accurately predicting future changes in ozone or global warming is our inability to predict the future evolution of the atmospheric concentrations of these gases. We need to understand the role of the biosphere in regulating the emissions of gases such as CH_4, CO_2, N_2O, and methyl chloride (CH_3C_l) to the atmosphere, and we need to know the most probable future industrial release rates of gases such as the CFC's, N_2O, CO, and CO_2 which depend upon economic, social, and political factors.

One important aspect of the ozone and global warming issues is that the atmospheric lifetimes of gases such as N_2O, CFC_{13}, and CF_2C_{12} are known to be very long. Therefore, if there is a change in atmospheric ozone or climate caused by increasing atmospheric concentrations of these gases the full recovery of the system will take several tens to hundreds of years after the emission of these gases into the atmosphere is ended. . . .

A Crucial Question

A crucial question is to assess the extent of changes in global ozone that have already taken place, and to compare the changes to what has been predicted by theory. The search for global ozone trends involves looking for small secular changes amid large natural variations that occur on many time scales. Observations of the total column content and the vertical distribution of ozone have been made for several decades using networks of different measurement techniques. Unfortunately, each of these observational techniques have certain limitations which tends to restrict our confidence in the results. These limitations arise from factors such as the lack of continuity of reliable calibration and the uneven geographic distribution of stations. Statistical analysis of the data is required to identify small trends, amongst high natural variability, using data from relatively few stations.

In general, analyses for the trends in the total global column content of ozone using data from the ground-based Dobson spectrophotometer network show no statistically significant trend since 1970, in agreement with model predictions for the same period when the changes in all of the trace gases are taken into account. It should be noted that the values of total global column ozone in the last three years have shown significant variability. Abnormally low values of total column ozone were observed in 1983 following the eruption of El-Chichon and the largest El-Nino event of this century. However, the values of total column ozone recovered in 1984, only to decrease significantly in 1985. . . .

Progress

It is clear that much progress has been made in our understanding of the physical and chemical processes that control the distribution of ozone. However, we must recognize that significant uncertainties in our knowledge remain, and that these can only be resolved by a vigorous program of research. It is essential that the US government and industry continue their strong commitment to studying the upper atmosphere, and that the scientific agencies continue their close collaboration at both the national and international level.

In summary, given what we know about the ozone and trace gas-chemistry-climate problem, we should recognize that we are conducting a global scale experiment on the Earth's atmosphere without a full understanding of the consequences.

THE OZONE LAYER

SERIOUS OZONE DEPLETION IS NOW OCCURRING

F. Sherwood Rowland

Dr. F. Sherwood Rowland is a professor of chemistry at the University of California in Irvine, California.

Points to Consider

1. How do chlorofluorocarbon gases harm the atmosphere?
2. How rapidly are these gases building up in the atmosphere?
3. How long can fluorocarbon gases last in the atmosphere?
4. What measured ozone losses have occurred over the Antarctic?
5. Why is this loss taking place and what should be done?

Excerpted from testimony by F. Sherwood Rowland before the Senate Subcommittee on Environmental Pollution, June, 1986.

Clearly, the most striking changes in ozone concentrations which have been observed since regular measurements began 60 years ago are the progressive major losses in the Antarctic during the spring season in the 1980s.

In June 1974 Dr. Mario Molina and I published our paper "Stratospheric Sink for Chlorofluoromethanes—Chlorine Atom Catalysed Destruction of Ozone" in the international scientific journal *Nature*. This paper carried the first outline of our hypothesis that the chlorofluorocarbon gases would eventually produce serious depletion of stratospheric ozone, and was summarized by the following abstract:

"Chlorofluoromethanes are being added to the environment in steadily increasing amounts. These compounds are chemically inert and may remain in the atmosphere for 40-150 years, and concentrations can be expected to reach 10-30 times present levels. Photodissociation of the chlorofluoromethanes in the stratosphere produces significant amounts of chlorine atoms, and leads to the destruction of stratospheric ozone."

These sentences can, with the benefit of 12 years of intensive study, now serve equally well as a brief summary of the facts of the chlorofluorocarbon-ozone problem.

Atmospheric Measurements

In early 1974, no measurements had yet been made of any chlorine-containing molecule anywhere in the stratosphere. Now, we have detailed evidence concerning at least ten chlorinated compounds in the stratosphere itself, and of many more throughout the lower atmosphere. The chlorofluoromethanes CCl_3F (Fluorocarbon-11) and CCl_2F_2 (Fluorocarbon-12) are, as they were designed to be, chemically inert. Molina and I reasoned that the usual processes which cleanse the atmosphere of chemical pollutants such as dissolution in raindrops or break-up by visible sunlight would not affect these compounds. The absence of effective removal processes for these chlorofluoromethanes led us to predict two important consequences: the average molecule of each would survive in the atmosphere unchanged for many decades; and removal would occur by destruction in the stratosphere after absorption of ultraviolet radiation.

"If you ask me, evolution is just another passing fad."

Both of these predictions have been fully confirmed by actual atmospheric measurements. Both of these compounds have been accumulating everywhere in the lower atmosphere at a very rapid rate and are now found at almost three times the concentrations measured in the early 1970s. This swift build-up is a clear indication that the average atmospheric lifetimes are very long, and current estimates place them at about 70 years for Fluorocarbon-11 and more than 100 years for Fluorocarbon-12. These lifetimes already make clear that any changes in the atmosphere caused by them will still be easily detected not only in the year 2000, but also in 2100 A.D. Although many other possible removal processes—often described as tropospheric sinks—have been proposed for these compounds, the facts in the atmosphere have shown these sinks to be totally unimportant.

The concentrations of Fluorocarbons-11 and -12, as well as those of another 10 or 12 halocarbon molecules, have been measured in flasks returned to the laboratory after being carried empty into the stratosphere by balloon and opened by remote control. . . .

Ozone Depletions

Wide variations have been recorded over the past decade in atmospheric model predictions of future total ozone depletions. Our initial 1974 estimate of eventual global depletion of ozone was a loss of 7-13%, one-hundred years or so in the future. Successive official reports from NASA and from the National Academy of Sciences have described the subsequent fluctuations in such predictions, with estimates which have more or less covered the range from 2% to 20%. The range from 2% to 20% was coincidentally the range of uncertainty attached to an estimate of 7% given in the first NAS report in 1976. . . .

However, throughout these large oscillations in the calculated total ozone loss, heavy losses have always been predicted for the upper stratosphere. The changes in the results for total ozone loss have been caused by large variations in the lower stratosphere, sometimes indicating decreases there and large global losses in total ozone, and sometimes indicating increases in the lower stratosphere and smaller total ozone loss. The same situation holds true today—the prediction of future ozone losses are much less certain for the lower stratosphere than for the upper. . . .

Major Ozone Losses

Clearly, the most striking changes in ozone concentrations which have been observed since regular measurements began 60 years ago are the progressive major losses in the Antarctic during the spring season in the 1980s. These losses were first reported by British scientists using data from ground stations in Antarctica, and have been verified and expanded by data from U.S. satellite instruments and from Japanese high altitude balloons. The Nimbus-7 satellite data for October 1985 show wide areas of the Antarctic continent with extremely low ozone values down to 140 Dobson units recorded by the British during the 1950s and 1960s. This loss of more than 50% in total ozone during the Antarctic spring is not only without precedent anywhere over the globe, but represents a decrease which has happened almost entirely during the 1980s. The initial British publication a year ago pointed out the remarkable correlation between this depletion in Antarctic ozone and the rapid growth in the atmospheric concentrations of the chlorofluoromethanes, and the connection is inescapable. Moreover, when either of the heterogeneous reactions of chlorine nitrate with water or hydrogen chloride is added to the existing two-dimensional atmospheric models, very large depletions of ozone are calculated for the Antarctic spring. A plausible chemical explanation now exists not only for the enormity of the losses, but also for why they occur over Antarctica, in the spring, and in the 1980s. . . .

Avoiding Regulation

The atmospheric models used in these debates have always indicated a very large range of uncertainty in the calculated predictions of eventual ozone loss. We could have adopted an attitude of prudent caution for the atmosphere and chosen courses of action based on the possibility that further experiments and study would confirm the largest predicted ozone depletions. These losses have always been judged not to be tolerable, and would have then required government regulatory actions placing severe limits on the total organochlorine concentration of the atmosphere, and therefore on the emissions of chlorofluorocarbons. The governments of the world have instead adopted an attitude of prudent caution toward interfering with the chlorofluorocarbon industry, have worried about the possibility that feedbacks and not-yet-discovered science would work to ameliorate or even eliminate the ozone problem, and have avoided severe regulations. With the exception of the bans in North America and Scandinavia on chlorofluorocarbons as propellants in aerosol sprays, no effective regulations exist. As a consequence, the world-wide usage of these compounds is once again expanding because of major increases in other uses, and development of new uses. The exploration of how to stimulate and then satisfy the demand for chlorofluorocarbons in the developing countries is also well under way.

Antarctic Ozone Hole

The Antarctic ozone hole has arrived as a profound shock, first because the losses of ozone are massive, and second because it was completely unpredicted. Instead of the unexpected working to ameliorate ozone depletion, it has produced huge losses. We are now in the position of having chosen to tolerate some unspecified amount of ozone depletion, and are now wondering how badly we have miscalculated. We now have a hole in the ozone layer which will last for a century or more, even if the entire world were to stop further emissions of chlorofluoro-carbons today—which is of course impossible.

Will the Antarctic hole deepen? Will it spread, and how soon, to other latitudes in both hemispheres? Can we afford to go for another 5 or 10 years of wait-and-see, of measuring, monitoring and studying?

If our prime concerns are the atmosphere, the ozone layer, and the people it shields, the obvious answer is to discontinue this experiment without waiting for all the answers.

THE OZONE LAYER

OZONE LOSS IS
NOT SIGNIFICANT

S. Robert Orfeo

Dr. S. Robert Orfeo is the chairman of the Fluorocarbon Program Panel of the Chemical Manufacturers Association.

Points to Consider

1. What are the objectives of the Fluorocarbon Program Panel?
2. How is the ozone loss over the Antarctic described?
3. What do studies show about atmospheric ozone loss?
4. What causes ozone loss in the stratosphere?
5. What actions should be taken to protect the ozone layer?

Excerpted from testimony by S. Robert Orfeo before the Senate Committee on Environment and Public Works, June, 1986.

Continued releases of CFCs will not pose a significant threat to the environment during the time required to gain a better understanding of the science.

The Fluorocarbon Program Panel (Panel) of the Chemical Manufacturers Association (CMA) welcomes the opportunity to address the issues of potential changes in atmospheric ozone levels and the greenhouse effect. The Panel represents most of the world's chlorofluorocarbon (CFC) producers. CFCs serve critical needs in refrigeration, air conditioning, and other diverse uses that are considered highly beneficial by society.

Global Concerns

The Fluorocarbon Program Panel shares the concern about atmospheric ozone levels and the greenhouse effect. The Panel believes that these global issues require international cooperation on research and monitoring. Based on this belief, the Panel began a research program 14 years ago to understand the potential effects of CFCs on the global environment. This program is coordinated with and complements research of government agencies in the United States and in other countries. The most recent review of atmospheric research programs shows that much progress has been made, but much remains to be learned. Based on research to date, in the judgment of the Panel, continued releases of CFCs will not pose a significant threat to the environment during the time required to gain a better understanding of the science. Continuation and augmentation of existing research programs will improve our understanding of the global environment and will provide a scientific basis for future courses of action.

The Program Panel

The Fluorocarbon Program Panel is an international group representing 19 CFC-producing companies from 10 countries. Its research program, administered by CMA, was initiated in 1972, to investigate the potential effects of CFCs on the environment. Over the years, the program has expanded its activities greatly. While its work remains focused primarily on the issue of potential changes in atmospheric ozone levels, it also includes work on the greenhouse effect. It has spent in excess of $18 million to date on research, and has had an annual budget of about $1.8

million in recent years. At least an equivalent amount has been spent by individual companies in support of this program. These figures do not include work conducted by individual companies on substitutes for CFCs. The Panel itself does not conduct any cooperative work in this area. . . .

Antarctic Spring

In addition, the Panel's research program addresses specific scientific questions. The most recent example is the work currently under way to understand the recently observed decrease in ozone in the Antarctic spring. A number of hypotheses have been proposed to explain this phenomenon; only some involve CFCs. Scientists agree that more data are needed to understand the observations, to test the hypotheses, and to establish a credible explanation. The Panel is cooperating with government agencies [in the United States: the National Aeronautical and Space Administration (NASA), the National Oceanic and Atmospheric Administration (NOAA), and the National Science Foundation (NSF)] in planning and funding campaigns of atmospheric measurements over Antarctica in 1986 and 1987. Simultaneously, the Panel is supporting laboratory studies, modeling projects, and data analysis programs to better understand the phenomenon. . . .

Long-Term Effects

The Panel has also funded the application of sophisticated statistical methods for analyzing long-term records of ozone data for possible trends. Again, this is a cooperative effort with researchers from government agencies and universities. The results of such trend analyses show that there has been no significant change in globally averaged total ozone. This finding is in agreement with model calculations. It is the total ozone level that controls the amount of ultraviolet radiation reaching the earth's surface. A recent analysis of a more limited and less reliable data set indicates that upper stratospheric ozone has decreased. This result is also in qualitative agreement with model calculations, but its significance is not yet understood.

Causes Unknown

While it is recognized that analysis of ozone data can identify an effect, it cannot establish cause. Furthermore, model calculations

Ozone Layer

A National Academy of Sciences study reduces the estimate of how much of the atmosphere's protective ozone layer is being destroyed by reactions with chlorine atoms in man-made chlorofluorocarbons.

The report from the academy's National Research Council said improved mathematical models and data from recent atmospheric research indicate human activities cause less change in atmospheric ozone than previously believed.

The research committee said continued release of chloroflurocarbon gases at current rates could reduce stratospheric ozone by between 2 percent and 4 percent by late in the next century.

In an earlier ozone report in 1979, the council estimated a possible ozone reduction in the range of 15 to 18 percent.

St. Paul Pioneer Press, February 23, 1984

suggest that there are much more sensitive indicators of potential stratospheric change than measurement of ozone itself. A critically needed program to monitor these indicators is being explored by NASA, NOAA, and the Panel, and should be supported. This program is an early detection network of ground-based monitoring stations. The main purposes of this network would be to: identify any changes in key atmospheric species; provide information on the causes of such changes; and most importantly, warn of future significant changes in ozone that may be induced by human activities well before they would actually occur. The feasibility of such a network was established at an international workshop held in Boulder, Colorado in March 1986. The workshop was sponsored jointly by NASA, NOAA and the Panel.

Initiating the early detection network, described at the Boulder workshop, would require a three part effort: adding a number of instruments at existing monitoring locations; establishing new sites; and continuing the current satellite and ground-based monitoring systems. Data from this network would provide additional constraints to test the models. It would also act as a calibration system for satellite-borne monitoring instruments, thereby resolving concerns about their intercomparability and calibration drift. The early detection system, coupled with the evidence that there has been no change in total ozone and the fact that model calculations show no change in total ozone for the next two to three decades, gives us confidence that time is available to conduct the research and monitoring necessary to establish the credibility of the long-range predictive capabilities of models. Based on the science, in the judgment of the Panel, there is no justification for additional regulations at this time.

Greenhouse Effect

The other issue under consideration is the greenhouse effect. CFCs are minor contributors to the greenhouse effect, the major contributor being carbon dioxide. To better understand the relationship of CFCs to the greenhouse effect, the Panel funds a two-dimensional (altitude and latitude) modeling effort. Two-dimensional models bridge the gap between simple one-dimensional (altitude only) models and much more realistic but very costly three-dimensional models. The magnitude of the total greenhouse effect is very uncertain due to our lack of understanding of factors such as clouds that can either decrease or increase the direct contribution of the many "greenhouse gases." The relative contribution of CFCs, however, can be calculated with reasonable certainty. At the present time, CFCs are calculated to contribute about 15% of the total effect. Future contributions of all trace gases will depend on their relative growth rates.

Government Programs

The existing U.S. Government programs have been instrumental in developing the current understanding of atmospheric sciences. These programs have advanced the understanding of processes controlling atmospheric ozone and climate. However, much more work is needed to quantify these processes and to determine the nature and the extent of the potential effects of man's activities on ozone and climate. Existing and planned programs should provide steady progress toward the goal of quantifying the theories of ozone and climate modification. These programs must continue to receive adequate funding if this goal is to be reached.

As scientific understanding has matured, the importance of long-term monitoring through an early detection network has been identified. Such a ground-based network, described earlier in this testimony, would serve several purposes. It would provide an early warning of significant changes in stratospheric ozone, provide additional data needed to improve model simulations of the atmosphere, and provide a reference set of measurements for comparison with satellite data. Establishing and maintaining the early detection network requires a long-term international funding commitment. The U.S. must continue to play a lead role in atmospheric sciences by not only maintaining existing and planned research and monitoring programs but also by fostering the implementation of this early detection network. This can be done by rapid deployment of network stations at U.S. sites, cooperating with scientists and agencies from other countries on development and deployment of stations on foreign sites, and providing adequate travel funds for government scientists to participate in this international program. This last issue of travel funds is a critical point that is essential to the success of the program.

Conclusion

Addressing long-range global scientific issues requires long-term international cooperation and commitment, which should be promoted through international bodies, such as the World Meteorological Organization and UNEP. The Panel, for its part, plans to continue its long-standing role in funding international research activities.

REGULATION OF POLLUTANTS IS NEEDED

Andrew Maguire

Andrew Maguire is the vice president of the World Resources Institute, an environmental organization in Washington, D.C.

Points to Consider

1. How will the choice of energy resources determine the history of the human species?
2. What international studies on climate are identified and what have they concluded?
3. How can ozone loss affect human populations?
4. What actions and programs are advocated to solve environmental problems?

Excerpted from testimony by Andrew Maguire before the Senate Subcommittee on Environmental Pollution, June, 1986.

We should adopt incentives and controls such as further reductions on the use of chlorofluoro-carbons and regulations that promote energy conservation.

From what we now know we can say quite literally that our human species will determine its destiny through our choice of energy sources, our controls on emissions of nitrous oxides and other atmospheric pollutants, our policies toward the use of chloro-fluorocarbons (CFCs), *and* the speed with which we act intelligently in these scientifically complex, interrelated areas.

For just these reasons, the scientific community is now encouraging policymakers to put climatic change high on the public agenda. Last October, an international conference of scientists gathered at Villach, Austria under the auspices of the UN Environment Programme (UNEP), the World Meterological Organization (WMO), and the International Council of Scientific Unions (ICSU) to assess the current status of knowledge concerning climate change. . . .

Scientific Perspective

Both the greenhouse effect and ozone modification are now accepted phenomena; we already have supporting empirical proof. The issue is not "whether" but "when", and with what consequences.

Several major international scientific reports have been published on these issues in the last six months—the Villach Conference, already described; the Department of Energy's four-volume "state-of-the-art" review of carbon dioxide buildup and its effects; and a detailed review of the chemical processes controlling concentrations of ozone and other climatically important trace gases produced by the WMO, NASA, and several other US and foreign agencies. These reviews reveal an impressive international consensus.

Especially striking are these specific conclusions:
1. A greenhouse warming of 3 to 8 degrees F is expected for a doubling of atmospheric CO_2 concentrations above pre-industrial levels. This is a global average that will be experienced through a warming larger nearer the poles than at the equator, with a possible sea level rise of 140 centimeters (4.5 feet).
2. Trace gases other than CO_2, including chlorofluorocarbons (CFCs), methane, and nitrous oxides, are now collectively

Preprinted by permission: Tribune Media Services

equal to CO_2 in their contribution to the greenhouse effect. These gases are also accumulating above the earth at faster rates. As a result, an effect equivalent to doubling CO_2 could occur at least twice as rapidly as previously thought: as early as the 2030s—less than 50 years.

3. Continued emissions of CFCs at current levels would result in ozone depletion (increasing incidence of fatal and non-fatal skin cancers and other serious health hazards, possibly individual suppression of the body's immune system) of 4 to 9 percent for constant CFC emissions and much greater reductions if emissions continue to grow. While growth in emissions of CO_2 and the other greenhouse gases could increase ozone in the lower stratosphere, partially offsetting dangerous depletion in the upper stratosphere, this would only occur in conjunction with a *more* serious greenhouse problem and potentially harmful changes resulting from shifting amounts of ozone at different altitudes and latitudes. . . .

Ozone Reductions

The effects of reductions in stratospheric ozone have received very little study but it is known that even small changes lead to thousands of new skin cancers. Harmful effects on the human immune system are also suspected as well as large economic losses due to accelerated damage to plastics and damage to some commercially valuable animals.

These changes, worrisome as they are, presume a gradual or largely linear process of change. In fact, scientists recognize that there may be critical points or even "mode switches" associated with changes in temperature and atmospheric trace gases that could lead to rapid, dramatic changes in climate. As the DOE state of the art report states, "The assumption that CO_2-climate change will be gradual and predictable is not unequivocally supported by evidence in the geologic record. . . The ice-core CO_2 record contains evidence for more rapid, concurrent changes in both CO_2 and climate on time scales on the order of a century. These past atmospheric CO_2 shifts are comparable to anthropogenic changes brought about since the early 19th century. . . .

Ozone Hole

The recent discovery of an "ozone hole" during springtime in the Antarctic provides a dramatic illustration of a change in the atmosphere occurring in a nonlinear fashion. Since the late 1970s, springtime ozone has dropped more than 40 percent. This change was not predicted and currently cannot be explained. The global increase in atmospheric chlorine due to emission of CFCs since the 1970s is thought by some but not all experts to be involved. Whatever the cause, such events give us further reason for caution with respect to the risks of conducting a giant global experiment in the atmosphere.

The lesson we draw from this brief scientific review is that conventional approaches to problem solving are inadequate to address what we are talking about today. For example, references to traditionally valuable methods of calculating "costs and benefits" may trivialize risks so all-encompassing. The greenhouse and ozone issues—and how we do or do not handle them—will affect all of humanity for generations to come.

Early Indicators

One challenge of greenhouse warming and ozone modification is that we must act before the dimensions of the problems are fully known or risk irreversible, catastrophic changes. Extended debates on control strategies and cost allocation schemes will

Sharp CFC Reductions

CFCs are the dominant chemicals in ozone depletion. In equivalent radiative terms, they also account for at least a sixth of the current atmospheric stock of greenhouse gases, and they are growing more rapidly than others. . . .

Given the gravity of both problems and the relative difficulty of as rapidly affecting emissions of other gases, CFC reductions clearly present the best present opportunity to stave off ozone depletion and to limit the global warming.

Natural Resources Defense Council, June 20, 1986

hardly be relevant once an unprecedented global warming has occurred. As William Ruckelshaus has said of the greenhouse problem, "The ultimate danger is that by remaining reliant on the 'catastrophe theory of planning' in an era producing catastrophes of a magnitude greater than in the past, we can place our institutions in situations where precipitate action is the sole option—and it is then that our institutions themselves can be imperilled and individual rights overrun."

First Steps

This hearing is momentous because it is the first to ask not only "what is the problem", but "what can we do?" It is tempting to throw up one's hands in despair upon realizing the full dimensions of these climate problems. They are truly global, and only an international response can be completely effective; substantial economic interests will be adversely affected; and the actions which are needed must precede a clear picture of the risks. But many modest steps are feasible and advisable in the near term, and, if taken, will make a difference in the outcome.

Of course, more research is urgently needed. The fact that the National Oceanic and Atmospheric Administration (NOAA) has still not committed to continued funding of a critical ozone-measuring satellite is a shocking example of the fragile support for even the most essential research on these issues. A major program on the role of the oceans in climate change ought to be another high priority. There is also very little support for research on the impacts of changes in climate and ozone. Despite the present budgetary pressures, the seriousness of the issues surely justifies funding at least double current levels.

However valuable, research alone is not the answer. Many of these issues will require decades to resolve. . . .

Conclusion

● "The human condition may depend on the ability of governments to implement policies to reduce the risks [of the greenhouse effect] before the full dimensions of these problems are understood, for by then it will almost surely be too late."
● "Both the greenhouse effect and ozone modification are now accepted phenomena; we already have supporting empirical proof. The issue is not 'whether,' but 'when,' and with what consequences."
● "From what we now know, we can say quite literally that our human species will determine its destiny through our choice of energy sources, our controls on emissions of nitrous oxides and other atmospheric pollutants, our policies toward the use of chlorofluorocarbons, *and* the speed with which we act intelligently in these scientifically complex, interrelated areas."
● "Conventional approaches to problem solving are inadequate to address what we are talking about today. References to traditionally valuable methods of calculating 'costs and benefits' may trivialize risks so all-encompassing."
● "We must place these issues much higher on the world's agenda. Planning by all government agencies should include explicit consideration of climate and ozone change."
● "The United States should exercise international leadership since we cannot solve these problems by ourselves. We should adopt incentives and controls that increase our options and buy time for solutions, such as further reductions on the use of chlorofluorocarbons and regulations that promote energy conservation."
● "If we curtail emissions of chlorofluorocarbons and growth in carbon dioxide-producing energy sources remains moderate, we can limit the extent of the effects, and delay the most serious changes for decades."

- "Policies to more directly limit carbon dioxide emissions may also soon be necessary. One logical direction is to consider a carbon tax, with funds raised going to plant trees and promote energy conservation."

THE OZONE LAYER

WE MUST NOT ACT
ON A THEORY

Alliance for Responsible CFC Policy

The Alliance for Responsible CFC Policy is a broad-based coalition of more than 500 users and producers of chlorofluoro-carbons (CFCs). The Alliance was formed in 1980 to bring a balanced perspective on CFC-related issues. Its main purpose is to ensure that government directives and policies pertaining to CFCs are scientifically sound and balanced, and are economically and socially effective.

Points to Consider

1. What are chlorofluorocarbons (CFCs) and how are they used?
2. What scientific theory has advocated the banning or use restrictions of CFCs?
3. How would life change without CFCs?
4. What actions should be taken to protect the environment?

Excerpted from a booklet by the Alliance for Responsible CFC Policy titled *Chlorofluorocarbons*, 1986.

There is ample time for the needed research, and any threat can be detected before any significant harm is done to the environment.

Chlorofluorocarbons (CFCs) are some of the most useful chemical compounds ever devised. They refrigerate our food; air condition our homes, workplaces, cars and public buildings; clean delicate electronic components; help insulate products; sterilize medical equipment and devices—the list is long and diverse.

Yet, CFCs may be banned or their manufacture and use restricted because of a theory.

The theory, first proposed in 1974, is that CFCs released into the atmosphere rise to the stratosphere, where they break down, releasing chlorine. In a complex chemical system, according to the theory, chlorine could reduce stratospheric ozone, causing an increase in the amount of harmful ultraviolet radiation reaching the earth.

For more than a decade, scientific studies supported by the industry and government have yielded useful information but have not determined that the theory is valid. In fact, actual measurements of stratospheric ozone over the last 20 years have detected no net depletion of the globally averaged ozone layer.

Nevertheless, the United States Environmental Protection Agency (EPA) prohibited the use of CFCs in nearly all aerosol products in 1978, and in 1980 proposed to consider further regulations that would place restrictions on all forms of CFCs. The Alliance for Responsible CFC Policy believes such restrictions would severely impact many industries and curtail the supply of beneficial—even essential—products. Moreover, they would force manufacturers to use substitutes for CFCs that are less efficient, more costly and potentially hazardous. In many cases, no suitable substitutes exist. The Alliance, a coalition of chlorofluorocarbon users and producers, believes further restrictions on CFCs at this time are unwarranted and would be against the public interest. The coalition supports continuing scientific investigations and international cooperation to determine whether CFCs do indeed pose a threat to the environment.

Scientific studies to date indicate the answer can be found and that the time required will not significantly increase the hazard—if such a hazard is found to exist.

107

The Alliance believes any further regulation of CFCs should be based on science, not supposition. . . .

Science vs. Supposition

The basic issue in the CFC/ozone controversy is whether a theoretical premise, the subject of continuing scientific research, is cause for further regulation. Secondary to this issue are questions of whether regulations are appropriate and practical, and what impact they would have on the public and the nation's economy.

EPA has continued to encourage regulation. Its 1978 ban on CFC aerosol propellants was followed in 1980 by announcement of the agency's intention to put a cap on all CFC production and then phase down production by as much as 70 percent.

Although EPA has not promulgated any further CFC regulation, the agency continues to single out CFCs in the ongoing stratospheric ozone risk-assessment program rather than more appropriately examining all substances that may affect the ozone layer. The importance of considering all substances that affect the ozone layer was pointed out in a recent report by the National Aeronautics and Space Administration (NASA) which said, "It is now recognized that the chemical effects of (CFC's, carbon monoxide, carbon dioxide, methane, nitrous oxide and the nitrogen oxides) on atmospheric ozone are strongly coupled and should not be considered in isolation."

EPA is also leading U.S. government efforts urging international CFC regulation through the United Nations Environment Programme.

The Alliance, while agreeing that the CFC/ozone theory raises serious questions that must be answered, nevertheless believes regulation should be based on facts, not theoretical assessments. It calls for a thorough, unbiased review of the real risks versus the real benefits of CFCs to society, as well as an assessment and understanding of how all substances that may affect the ozone layer interact.

Scientific investigations are going on which will lead to identification and measurement of the risks, if they exist. The results will warn of any environmental threat in ample time to limit any significant adverse effects.

Worldwide Action

It was also our view that further unilateral regulation of CFCs by the United States, without worldwide coordination and similar action, would only serve to penalize economic growth in the United States, injure U.S. industries to the benefit of our international competition, and provide little, if any, environmental protection.

Richard Barnett, chairman of the Alliance for Responsible CFC Policy, March 7, 1986

The Alliance sees no need to rush into regulatory decisions until the facts are known.

Why Not Switch to Other Compounds?

For many CFC applications, there are possible substitutes. Indeed, some substitutes are being used today. Almost without exception, however, the substitutes do not have all the desirable characteristics of CFCs, and they often have such undesirable properties as flammability, explosiveness, toxicity or corrosiveness—sometimes a combination of all four.

Could better compounds be created? Possibly, but so far efforts to do so have failed. Since the 1978 United States ban on CFC propellants, for example, several CFC producers have worked long and hard to find substitute compounds. All efforts have fallen short. There is a finite number of chemical building blocks. More complex molecules can be formed, of course, but their very complexity gives them undesirable properties not present in CFCs.

From all indications, the CFC family is unique.

Life Wouldn't be the Same Without Them

By any measurement, CFCs are important to modern life:
● They help us meet basic needs—food, shelter, health care, communications, leisure, transportation.

- They contribute immeasurably to our comfort, safety and productivity.
- In the United States alone, CFCs are used by some 5,000 businesses at nearly 375,000 locations to produce goods and services worth more than $28 billion a year. CFC-related jobs total 715,000. . . .

Searching for Answers

Ever since the question of whether CFCs damage the ozone layer was raised in 1974, science, industry and government have engaged in intensive efforts to determine the facts. In all this time, CFC emissions have not been found to have perturbed the natural behavior of ozone. Indeed, sophisticated scientific analysis has shown no persistent change in the total amount of atmospheric ozone over the last 20 years.

The research has focused primarily on the highly complex chemistry of the stratosphere. It is known that ozone acts as a giant filter to screen out some of the sun's harmful ultraviolet rays. The belief is that depletion of the ozone layer could result in increased incidence of some forms of skin cancer and damage to certain food crops and aquatic life.

Studying the stratosphere is extremely difficult, not only because of its distance from the earth (from eight to 30 miles) but also because the concentrations of ozone it contains are subject to frequent and often large natural fluctuations. However, by analysis of samples taken at various places and times through computer modeling, a great deal has been learned about stratospheric chemistry.

It appears that the amount and distribution of ozone in the atmosphere are maintained by a dynamic balance between production (from solar ultraviolet radiation), destruction (by radicals derived from several trace gases) and transport by atmospheric motion. The process is not completely understood.

It is evident, however, that industrial, agricultural and natural processes play a part in production of the trace gases. For example, carbon dioxide is increasing in the atmosphere due to increased burning of fossil fuels. Methane levels are also rising from sources thought to be natural wetlands, rice paddies and fermentation processes in cattle and other ruminants.

A 1986 report to Congress by the National Aeronautics and Space Administration (NASA) describes the current status of atmospheric science: what has been learned, what remains scientifically uncertain and what research still needs to be done.

A central point in the report is that the chemical effects of trace substances on atmospheric ozone are strongly coupled and should not be studied in isolation.

The report points out that low ozone levels observed in the winter of 1982-83 appear to be due to natural, rather than man-made causes. A recently observed ozone decrease in the Antarctic is not yet understood, the report states, and further study is needed to determine whether it is significant.

Though atmospheric observations have established the presence of key constituents, the report points out that current computer models do not adequately reproduce the present-day atmosphere. Discrepancies between observations and calculations limit the scientists' confidence in the predictive capability of the models. . . .

In Summary

Chlorofluorocarbons, because of their utility, safety and unusual combination of desirable properties, have become important in nearly every aspect of modern life. To restrict their production and use would be detrimental to the public interest.

The Alliance for Responsible CFC Policy believes:

- In the absence of more compelling scientific evidence that CFCs present a hazard to the environment, further regulation is unwarranted at this time.
- Research into the scientific questions raised by the ozone depletion and "greenhouse effect" theories should be diligently pursued by the industry and government.
- There is ample time for the needed research, and any threat can be detected before any significant harm is done to the environment.
- Research should include studies of other substances that may affect the stratosphere, not just CFCs.
- Since computer model calculations of the theoretical effects of CFCs on ozone have been inconclusive and everchanging, they do not constitute a sound basis for regulation. Actual measurements in the stratosphere are also a necessary part of the decision-making process.
- Ozone depletion, if it occurs, would be a global issue, so international cooperation should be enlisted for the scientific studies and assessments. If the studies indicate regulatory action

is needed, it should be taken in concert with other nations, not unilaterally.

- If scientific findings should prove regulation of CFCs is necessary, the regulations should be based on a balanced assessment of risks and benefits from CFC use, with due consideration given to risks from the use of known alternative compounds.

EXAMINING COUNTERPOINTS

This activity may be used as an individual study guide for students in libraries and resource centers or as a discussion catalyst in small group and classroom discussions.

Guidelines

Social issues are usually complex, but often problems become oversimplified in political debates and discussion. Usually a polarized version of social conflict does not adequately represent the diversity of views that surround social conflicts.

Examine the counterpoints below. Then write down other possible interpretations of the issue than the two arguments stated in the counterpoints which follow.

CFCs: Sharp Reductions Are Necessary: The Point

CFCs are the dominant chemicals in ozone depletion. In equivalent radiative terms, they also account for at least a sixth of the current atmospheric stock of greenhouse gases, and they are growing more rapidly than others. But there are now compelling reasons to conclude that it is no longer sufficient to talk just about avoiding increases in CFC production. Even with such a cap, CFC levels reaching the stratosphere will continue to increase. Given the gravity of both problems, CFC reductions clearly present the best present opportunity to stave off ozone depletion and to limit the global warming. (National Resource Defense Council, 1986)

CFC Reductions Are Not Necessary: The Counterpoint

The CFC industries in the United States oppose CFC regulations as being premature and unwarranted. Research, not further regulation of chemicals essential to society, should be the focus of ozone protection at this time.

U.S. CFC users and producers are concerned that the U.S. government has proposed actions that clearly go beyond any domestic regulatory measures adopted in the U.S.

Discussion about possible inadvertent modification of the ozone layer has assumed worldwide dimensions. Actions by one nation alone, even the United States, to limit or control the uses of one or more of the substances that may modify ozone would not alleviate a problem, should it be found that one is developing.

No reduction of globally-averaged ozone levels has been observed to date that can be attributed to any single chemical or combination of chemicals—natural or manmade. (Alliance For Responsible CFC Policy, 1986)

CHAPTER 4

THE NEED FOR ACTION?

THE NEED FOR ACTION?

A PLAN TO SAVE
THE PLANET

John H. Chafee

John H. Chafee is a Republican United States Senator from Vermont and a leading spokesman in the Senate for environmental protection.

Points to Consider

1. What does scientific evidence say about ozone depletion and the greenhouse effect?
2. How should we deal with questions of uncertainty?
3. What choices do we have today?
4. What immediate actions and initiatives should be taken?

Excerpted from a statement by Senator John H. Chafee before the Senate Subcommittee on Environmental Pollution, June, 1986.

"We cannot afford to give chemicals the same constitutional rights that we enjoy under the law . . . chemicals are not innocent until proven guilty".

There is a very real possibility that man—through ignorance or indifference, or both—is irreversibly altering the ability of our atmosphere to perform basic life support functions for the planet.

Ozone depletion and the greenhouse effect can no longer be treated solely as important scientific questions. They must be seen as critical problems facing the nations of the world. These are problems that demand solutions.

A recent assessment of the greenhouse problem by the Department of Energy summed up the stakes in stark terms:

"Human effects on atmospheric composition and the size and operations of the terrestrial ecosystems may yet overwhelm the life support system crafted in nature over billions of years."

This is not a matter of Chicken Little telling us the sky is falling. The scientific evidence, some of which we will hear today, is telling us we have a problem, a serious problem. There is much that we know. There is a great deal that we can predict with a fair amount of certainty.

Uncertainty

Now, it is true that we lack the tools to close all of the scientific gaps. We don't completely understand our climate systems and we cannot predict precise outcomes. But we will always be faced with a level of uncertainty. It is a fact that the current gaps in scientific knowledge may not be closed for many years. Therefore, the question raised is this: Can we continue to risk so much when we do not know the detailed nature of the outcome?

To my mind, the risks are so great that we must avoid continuing on a path that will irreversibly alter our environment *unless* we know that it is safe to proceed down that path. Scientists have characterized our treatment of the greenhouse effect as a global experiment. It strikes me as a form of planetary Russian roulette.

We should not be experimenting with the Earth's life support systems until we know that—when the experiment is concluded—the results will be benign. As Russell Peterson, former chairman of the President's Council on Environmental Quality who worked as a chemist for twenty-six years has said, "We cannot afford to give chemicals the same constitutional rights that we enjoy under the law . . . chemicals are not innocent until proven guilty".

Policy Choices

By not making policy choices today, by sticking to a "wait and see" approach, we may in fact be making a passive choice. By allowing these gases to continue to build up in the atmosphere, this generation may be committing all of us to severe economic and environmental disruption without ever having decided that the value of "business as usual" is worth the risks.

Those who believe that these are problems to be dealt with by future generations are misleading themselves. Man's activities to date may have already committed us to some level of temperature change. If historical evidence is any guide, a slight warming may be enough to turn productive, temperate climates into deserts. To quote from another recent Department of Energy report, "large changes in both precipitation and the extent of deserts and grasslands can be associated with relatively small variations in the global mean temperature."

The path that society is following today is much like driving a car towards the edge of a cliff. We have a choice. We can go ahead, take no action and drive off the edge—figuring that, since the car will not hit the bottom of the canyon until our generation is already long gone, the problem of coping with what we have made inevitable, is for future generations to deal with. We can hope that they will learn how to adapt. On the other hand, we can put the brakes on now, before the car gets any closer to the edge of the cliff and before we reach a point where momentum will take us over the edge, with or without application of the brakes.

Needed Action

Having painted a bleak picture, the question arises: What do we do about all of this?

The first thing we should do is to ratify the Vienna Convention for the Protection of the Ozone Layer which is pending on the Senate calendar. The Convention is the first world-wide legal instrument directed to the protection of the atmosphere as a resource. It has the support of industry, environmental groups and the Administration and should be ratified as soon as possible.

Next, it is important to focus attention on the potential effects of ozone depletion and of climate change and on the choices that we as a global society must make if we are to avoid further build-up of harmful gases in the atmosphere. These are no longer just science issues. They are now policy issues.

As evidenced by the October 1985 Villach Conference, there is now an international consensus among the scientific community. Although there will always be dissenters, those who claim, for example, that the Earth is actually cooling not warming, the scientific community has told us with unusual clarity that we have a problem. We must not allow their message to fall on deaf ears.

New Initiatives

To move us along in the right direction, I intend to ask my colleagues to join me in six new initiatives.

First, we will be asking EPA Administrator Thomas and the Executive Director of the Office of Technology Assessment, Jack Gibbons, to launch immediate and separate studies setting forth policy options that, if implemented, would stabilize the levels of atmospheric gases. These studies are expected to address significant changes in enery—in terms of both improvements in energy efficiency and development of alternatives to fossil fuels—, reductions in the use of CFCs, ways to reduce other greenhouse gases such as methane and nitrous oxides, as well as rates of deforestation and reforestation efforts. The thought is not to embark on a five or ten year study but to conclude these studies in fairly short order, say one or two years.

Second, the National Academy of Sciences will be asked to review existing gaps in scientific knowledge and to make recommendations to us on how to close these gaps. Again, this is a task that should take several months, not several years. In developing recommendations, the Academy will be expected to consult with all of the relevant federal agencies.

Third, the Department of State should make its best effort to bring these issues to the attention of other nations. At a minimum, that effort should include discussions at the next summit meeting with the Soviet Union and at the next international economic summit meeting. The fact that the Soviet Union contains 44% of the world's coal reserves makes their involvement particularly important. Similarly, the vast coal reserves in the People's Republic of China makes them major players in this matter.

Fourth, we will be urging the United Nations Environment Program and the WMO (World Meteorological Organization) to expand their efforts to assess climate change problems and its social and economic impacts. As a follow-up to the successful Villach Conference of October 1985, we need these organizations to look at social and economic *effects* of climate change and *policy options* to reduce atmospheric pollution. Such an assessment should enable UNEP to convene a meeting to negotiate a convention on climate change in the near future.

Fifth, EPA will also be asked to coordinate a study on the environmental effects of climate change. This study should be designed to solicit the opinions of knowledgeable people throughout the country through a process that includes public hearings and meetings.

And finally, we will be asking the President's Council on Environmental Quality to issue a directive to all federal agencies to recognize ozone depletion, the greenhouse effect and climate

120

Fossil Fuel

The biggest difficulty is how to grapple with it. The political climate is such that people aren't willing to look down the road in Congress right now.

What's needed is an energy bill that links carbon dioxide with energy conservation. "In essence, we have to get ourselves off fossil fuel."

Jeffrey Knight, Friends of the Earth, June 1981

change as environmental impacts that must be considered.

Conclusion

It seems that the problems man creates for our planet are never ending. But we have found solutions for prior difficulties and we will for these as well. What is required is for all of us to do a better job of *anticipating* and responding to today's new environmental warnings *before* they become tomorrow's environmental tragedies.

THE NEED FOR ACTION?

WE ALREADY HAVE A
PLAN OF ACTION

Lee M. Thomas

*Lee M. Thomas is the administrator of the U.S. Environmental
Protection Agency (EPA).*

Points to Consider

1. How significant is the present threat to our atmosphere?
2. What restrictive action has the United States already taken
   in the area on banning chemical releases in the atmosphere?
3. What research on climate change has EPA sponsored?

Excerpted from testimony by Lee M. Thomas before the Senate
Committee on Environment and Public Works, June, 1986.

The U.S. has already taken a leadership positio
by banning non-essential uses of aerosols.

Although we at EPA usually focus on pollution that directly affects land, water, and the air we breathe, we must not ignore the environmental significance of changes now occurring in the composition of the earth's atmosphere from our industrial activities.

Global Environment

Our atmosphere plays a fundamental role in shaping and protecting our planet's environment. Changes in its delicate chemical and physical balance could possibly lead to two separate but related problems.

First, the ozone layer of our stratosphere currently protects us from exposure to most of the sun's damaging ultraviolet radiation. Partial depletion of the ozone layer would increase our exposure to the potentially damaging part of the solar spectrum, leading to adverse health and environmental effects.

Second, certain gases in the atmosphere, known as greenhouse gases, form a "thermal blanket" around the earth by blocking part of the infrared radiation reflected from the earth's surface. The presence of these greenhouse gases act to maintain our planet's current moderate temperature. Increases in the amount of these greenhouse gases would result in a rise in the earth's average temperature. . . .

Let me discuss what I believe to be some of the unique factors associated with both stratospheric ozone depletion and climate change.

Both issues are clear examples of a "global commons" environmental problem. All nations are responsible for contributing to recent changes in our atmosphere—although the industrially developed nations must shoulder more of the responsibility. However, all nations would be affected by depletion of the ozone layer or by global climate changes. Therefore, the international community will need to cooperate in any effective solution to these problems. The U.S. has already taken a leadership position by banning non-essential uses of aerosols. . . .

Given the scientific uncertainties, we recognize that any action taken now has a cost associated with it which, as we learn more, may prove unwarranted. Thus, any analysis of whether actions to slow emissions are necessary must compare the costs and risks of acting now or acting later. . . .

Ozone Depletion

On January 10, 1986, EPA published in the *Federal Register* notice of an expanded program to meet our responsibilities under the Clean Air Act. While the program does not commit EPA to taking specific actions to further regulate CFCs or other ozone modifying gases, it does initiate EPA's rulemaking process which will provide the basis for a decision concerning whether additional controls are warranted. The plan commits us to making a final decision by November 1, 1987. The Agency is under a court order to make a decision by that time.

The Agency's Stratospheric Protection Plan contains both a domestic and international focus.

On the Domestic Front:

● We are committed to a final decision concerning the need for additional regulations by November 1987;

● We are re-establishing the Interagency Coordinating Committee on Stratospheric Ozone Protection to ensure proper coordination of research across all federal agencies;

● We have held one domestic workshop on economic aspects of this issue, and have a second scheduled for late July.

● We are calculating the effects of UV-B on agricultural plants and aquatic organisms, and have a small effort on human health effects.

On the International Front:

● We actively supported the Vienna Convention for the Protection of the Ozone Layer, pending before the Senate for its advice and consent. This Convention provides for international cooperation and support for scientific research on the stratospheric ozone issue.

● We are examining a full range of international strategies as part of our risk management/risk assessment process and will decide our negotiating position on that basis when international negotiations resume in the fall. Consequently, we have stepped back from our previous unsuccessful efforts to persuade other nations to follow our lead and ban the use of chlorofluorocarbons in nonessential aerosol products.

124

U.S. Government Accomplishments

About one decade ago leading scientists concerned about the CO_2 issue prepared a number of recommendations which in large measure have now been implemented. One recommendation called for reliable CO_2 standards. Accurate and reliable measurements of atmospheric CO_2 depend on high-quality standards. I am pleased to report that the National Bureau of Standards has now prepared stable standards for use by the international CO_2 measurement community. Six standard reference gases have been prepared and certified.

Alvin W. Trivelpiece, U.S. Department of Energy, June 11, 1986

- We are actively participating in the United Nations Environment Programme (UNEP) workshop on economic issues related to ozone protection and are co-sponsors of an international conference on Health and Environment Effects here in Washington next week
- We are moving in step with the timetable established by UNEP so that we will have an adequate information base for deciding our international and domestic positions.

We recognize that several important factors must not be overlooked. To the extent that action is needed, it is essential that the international community move forward to deal with these issues together. Further, we realize that our analysis must include all trace gases that may modify the ozone layer and not just CFCs. In the case of CFCs, we recognize that they are extremely important chemicals used across a broad spectrum of industrial and consumer goods. For some uses, we recognize that no effective alternative chemical currently exists.

Finally, we recognize that the potential risks we face are generally long-term and that, if action proves warranted, any regulatory approach selected should be structured in a way which minimizes costs and disruption to producers and users.

In summary, I feel that our recently initiated stratospheric protection plan provides a comprehensive basis for us to move forward to evaluate the need for additional controls consistent with our duties under the Clean Air Act.

Climate Change Activities

As I discussed earlier in this statement, the potential environmental implications of the rate and magnitude of global warming now predicted by climate models can only begin to be assessed. Yet, without this knowledge of the possible implications of continuing to add greenhouse gases to our atmosphere, any decision on future policy action would be premature.

Because several other agencies have extensive and quite excellent research programs aimed at understanding important aspects of this problem, I am going to limit my remarks specifically to those areas where EPA has sought to contribute.

Over the past four years we have supported a small, but active program focused on:

● evaluating likely future trends in emissions on non-CO_2 greenhouse gases (e.g., chlorofluorocarbons, nitrous oxides, methane);

● developing climate change scenarios (e.g., changes in temperature, water availability, and sea level) that could be used by our researchers and others to estimate possible economic and environmental effects from such climate changes; and

● working with outside groups to better understand the potential effects of climate change scenarios on such activities as forest productivity (with National Forests Products Association and Conservation Foundation); electric utility planning (with EPRI and EEI); salinity of drinking water (Delaware River Basin Commission).

Through these case studies, we hope to begin to understand the possible economic and environmental implications of climate change.

To augment our current efforts, I have recently established an EPA Climate Change Working Group that cuts across several offices within the Agency. This group will report to me on what additional activities, if any, might be initiated to effectively deal with this issue in a timely manner. I expect that we will focus our resources on the following areas of policy and research needs:

● estimating trends in greenhouse gas emissions and determining possible control options;

- developing and evaluating improved scenarios to be used for estimating potential environmental effects of trace gases; and
- expanding long-term research focused on possible environmental effects.

In addition to specific research and analysis, we also intend to work through the Interagency National Climate Policy Board and Program to ensure that the entire federal research effort in this area is responsive to environmental concerns, and to expand our international efforts related to climate change. While supporting the recent Villach Conference statement on climate change, we fully recognize that increased understanding of this issue is essential before any international regulatory action will be possible.

To conclude, I would like to emphasize both my deep concern about the potential environmental risks associated with these issues and the complexity of developing a response which effectively responds to their unique characteristics. Despite these complexities, I want to assure you that we are moving forward in a timely manner to responsibly deal with these issues.

INTERPRETING EDITORIAL CARTOONS

This activity may be used as an individualized study guide for students in libraries and resource centers or as a discussion catalyst in small group and classroom discussions.

Although cartoons are usually humorous, the main intent of most political cartoonists is not to entertain. Cartoons express serious social comment about important issues. Using graphic

'Will the one thousand forty-seventh meeting of the two millionth Commission to Study Acid Rain please come to order!'

By Bill Sanders, *Milwaukee Journal,* Permission of News America Syndicate and visual arts, the cartoonist expresses opinions and attitudes. By employing an entertaining and often light-hearted visual format, cartoonists may have as much or more impact on national and world issues as editorial and syndicated columnists.

Points to Consider

1. Examine the two cartoons in this activity.

2. How would you describe the message of each cartoon? Try to describe each message in one to three sentences.

3. Do you agree with the message expressed in either cartoon? Why or why not?

4. Do either of the cartoons support the author's point of view in any of the readings in this book? If the answer is yes, be specific about which reading or readings and why.

5. Are any of the readings in chapter four in basic agreement with either of the cartoons?

BIBLIOGRAPHY

Albanese, A. and M. Steinberg, (1980), **Environmental Control Technology for Atmospheric Carbon Dioxide,** Brookhaven National Lab, prepared for the U.S. Dept. of Energy, New York, N.Y.

Anderer, J., A. McDonald, and N. Nakicenovic, (1980), **Energy in a Finite World,** International Institute of Applied Systems Analysis, Ballinger Publishing Co., Cambridge, Mass.

Arrhenius, S., (1896), "On the Influence of Carbonic Acid in the Air upon the Temperature of the Ground," **Philos Mag., 41,** 237.

Bach, W., J. Pankrath, and J. Williams (Ed.), (1980), **Interactions of Energy and Climate,** Reidel Publishing Co., Boston, Mass.

Baes, Jr., C.F., S.E. Beall, D.W. Lee, and G. Marland, (1980), "The Collection, Disposal, and Storage of Carbon Dioxide," in W. Bach et al. (Eds.), **Interactions of Energy and Climate,** Reidel Publishing Corp., Boston, Mass.

Barnola, J.M., D. Reynaud, A. Neftel, and H. Oeschger, (1983), "Comparison of CO_2 Measurements by Two Laboratories on Air from Bubbles in Polar Ice," **Nature, 303,** 410.

Bell, P.R., (1982), "Methane Hydrate and the Carbon Dioxide Question," in Clarke (Ed.), **Carbon Dioxide Review 1982,** Clarendon Press, New York.

Bindschadler, R., (1985). Contribution of the Greenland Ice Cap to Changing Sea Level. In M. F. Meier, (1985). **Glaciers Ice Sheets and Sea Level.** Washington, D.C.: National Academy Press.

Brewbaker, J. (Ed.), (1980), "Giant Leucaena (Koa Haole) Energy Tree Farm," Hawaii Natural Energy Institute, Hawaii.

Broecker, W.S., P. Tsung-Hung, and R. Engh, (1980), "Modeling the Carbon System," **Radiocarbon, 22,** 565.

Broecker, W.S., J.H. Nuckolls, P.S. Connell, and J. Chang, (1983), "CO_2: Backstop Against a Bad CO_2 Trip?" unpublished draft.

Budyko, M.I., (1969), "The Effect of Solar Radiation Variations on the Climate of the Earth," **Tellus, 21,** 611.

Budyko, M.I., (1974), "The Method of Climate Modification," (in Russian), **Meterology and Hydrology, 2,** 91.

Callender, G., (1939), "The Artificial Production of Carbon Dioxide and its Influence on Temperature," **Quarterly Journal of Royal Meteorological Society, 64,** 223.

Chamberlain, J., H. Foley, G. MacDonald, and M. Ruderman, (1982), "Climate Effects of Minor Atmospheric Constituents," in Clarke, (Ed.), **Carbon Dioxide Review: 1982,** Clarendon Press, New York, N.Y.

Charney, J., (1979), **Carbon Dioxide and Climate: A Scientific Assessment,** National Academy of Science, National Academy Press, Washington, D.C.

Clark, W. (Ed.), (1982), **Carbon Dioxide Review 1982,** Clarendon Press, New York, N.Y.

Council on Environmental Quality, (1981), **Global Energy Futures and the Carbon Dioxide Problem,** U.S. Government Printing Office, Washington, D.C.

Donner L., and Ramanathan V., (1980), "Methane and Nitrous Oxide: Their Effects on Terrestrial Climate," **Journal of Atmospheric Science, 37,** 119.

Dyson, F., (1976), "Can We Control the Amount of Carbon Dioxide in the Atmosphere?" Institute for Energy Analysis, Oak Ridge, Tenn.

Emanuel, W., G. Kilbugh, and J. Olson, (1981), "Modeling the Circulation of Carbon in the World's Terrestrial Ecosystems," in B. Bolin, et al. (Eds.), **Modeling the Global Carbon—Scope 16,** John Wiley, N.Y.

Edmonds, J. and J. Reilly, (1983a), "A Long Term, Global, Energy, Economic Model of Carbon Dioxide Release From Fossil Fuel Use," **Energy Economics, 5,** 74.

Edmonds, J. and J. Reilly, (1983b), "Global Energy Production and Use to the Year 2050," **Energy, 8,** 419.

Edmonds, J., J. Reilly, and R. Doughery, (1981), "Determinants of Energy Demand to 2050," Institute for Energy Analysis, Washington, D.C.

Flohn, H., (1981), **Life on a Warmer Earth,** International Institute for Applied Systems Analysis, Laxenburg, Austria.

Gilland, R., (1982), "Solar, Volcanic, and CO_2 Forcing of Recent Climatic Changes," **Climate Change, 4,** 111.

Greenburg, D., (1982), "Sequestering," prepared for the Office of Policy Analysis, U.S. Environmental Protection Agency, Washington, D.C.

Hansen, J., D. Johnson, A. Lacis, S. Lebedeff, P. Lee, D. Rind, and G. Russell, (1981), "Climate Impacts of Increasing Atmospheric CO_2," **Science, 213,** 957.

Hansen, J., G. Russell, D. Rind, P. Stone, A. Lacis, S. Lebedeff, R. Ruedy, and L. Travis, (1983), "Efficient Three Dimensional Global Models for Climate Studies: Models I and II," **Monthly Weather Review, 111,** 609.

Hilts, P.J., (1983), "El Nino Weather Disasters Continue," **Washington Post,** June 14, 1.

Hoffman, J., (1983), "Projecting Future Sea Level: Methodology, Estimates to the Year 2100, and Research Needs," U.S. Environmental Protection Agency, Washington, D.C.

ICF Inc., (1983), "Surrogate Measures for Assessing the Economic Impacts of Carbon Dioxide Mitigation," prepared for The Office of Policy Analysis, U.S. Environmental Protection Agency, Washington, D.C.

Keeling, C.D., R. Bacastow, and T. Whorf, (1982), "Measurements of The Concentrations of Carbon Dioxide at Mauna Loa Observatory, Hawaii," in Clarke, (Ed.), **Carbon Dioxide Review: 1982,** Clarendon Press, New York, N.Y.

Kellogg W. and R. Schware, (1981), **Climate Change and Society,** Westview Press, Boulder, Colorado.

Kellogg, W., (1979), "Influence of Mankind on Climate," **Annual Review of Earth Planetary Science, 7,** 63.

Lacis, A., J. Hansen, P. Lee, T. Mitchell, and S. Lebedeff, (1981), "Greenhouse Effect of Trace Gases, 1970-80," **Geophys. Res. Letters, 8,** 1035.

Lave, L., (1981), "The Carbon-Dioxide Problem: A More Feasible Social Response," **Technology Review, 84,** 22.

Leach, G., (1976), **Energy and Food Production,** IPC Science and Technology Press, Guildford, England.

Lovins, A., L. Lovins, F. Krause, and W. Bach, (1981) **Least-Cost Energy,** Brick House Publishing Co., Andover, Mass.

MacDonald, G., (Ed.), (1982), **The Long-Term Impacts of Increasing Atmospheric Carbon Dioxide Levels,** Ballinger Publishing Co., Cambridge, Mass.

Manabe, S. and R. Wetherald, (1980), "On the Distribution of Climate Change Resulting from an Increase in CO_2 Content of the Atmosphere," **Journal of Atmospheric Sciences, 37,** 99.

Marchetti, C., (1977), "On Geoengineering and the CO_2 Problem," **Climate Change, 1,** 59.

Meier, M.F. et al., (1985), **Glaciers, Ice Sheets and Sea Level: Effect of a CO_2-Induced Climatic Change,** Washington, D.C.: National Academy Press.

Mitchell, J. Murray, (1977), "Some Considerations of Climatic Variability in the Context of Future CO_2 Effects on Global-Scale Climate" in Elliott and Machta (Eds.) **Carbon Dioxide Effects Research and Assessment Program: Workshop on the Global Effects of Carbon Dioxide From Fossil Fuels,** March 7-11, 1977. U.S. Government Printing Office, Washington, D.C.

Mork, K., (1981), **Energy Prices, Inflation, and Economic Activity,** Ballinger Publishing Co., Cambridge, Mass.

NASA, (1983), "The Climate Impact of Increasing Atmospheric CO_2 with Emphasis on Water Availability and Hydrology in the United States," Draft, prepared for the U.S. Environmental Protection Agency by the Goddard Institute for Space Studies, New York, N.Y.

Nordhaus, W., (1977), "Strategies for the Control of Carbon Dioxide," Cowles Foundation Discussion Paper No. 443, New Haven, Conn.

Olson, J., (1982), "Earth's Vegetation and Atmospheric Carbon Dioxide," in Clarke, (Ed.), **Carbon Dioxide Review: 1982,** Claredon Press, New York, N.Y.

Oeschger, H., and M. Heimann, (1983), "Uncertainties of Predictions of Future Atmospheric CO_2 Concentrations," **Journal of Geophysical Research, 88,** 1258.

Ramanathan, V., H.B. Singh, R.J. Cicerone, and J.T. Kiehl, (1985), Trace Gas Trends and Their Potential Role in Climate Change, **Journal of Geophysical Research** (August)

Rasmussen, R. and M. Khalil, (1981), "Increase in Concentration of Atmospheric Methane," **Atmospheric Environment, 15,** 883.

Reilly, J., R. Dougher, and J. Edmonds, (1981), **Determinants of Global Energy Supply to the Year 2050,** Institute for Energy Analysis, Washington, D.C.

Rotty, R. M., (1983), "Distribution of and Changes in Industrial Carbon Dioxide Production," **Journal of Geophysical Research, 88,** 1301.

Scroggin, D. and R. Harris, (1981), "The Carbon-Dioxide Problem: Reduction at the Source," **Technology Review, 84,** 22.

Seidel, H.F., (1985), Water Utility Operating Data: An Analysis, **Journal American Water Works Association,** 77(5):34-41

Smagorinsky, J., (Chairman), (1982), **Carbon Dioxide and Climate: A Second Assessment,** National Academy of Sciences, National Academy Press, Washington, D.C.

Titus, J.G., (1985), Sea Level Rise and Wetlands Loss in **Coastal Zone '85** edited by O.T. Magoon, H. Converse, D. Miner, D. Clark and L.T. Tobin, New York: American Society of Civil Engineers.

United Nations Environment Program and World Meterological Organization (UNEP), (1985), International Assessment of the Role of Carbon Dioxide and of Other Greenhouse Gases in Climate Variations and Associated Impacts, Conference Statement, Geneva: United Nations Environment Program.

U.S. Dept. of Interior, (1980), **Minerals Yearbook,** prepared by the Bureau of Mines, U.S. Government Printing Office, Washington, D.C.

U.S. Dept. of Energy, (1981), "International Energy Indicators," prepared by International Affairs Office of Market Analysis, U.S. DOE, Washington, D.C., DOE/A-0010/7.

Wang, W. and J. Pinto, (1980), "Climate Effects due to Halogenated Compounds in the Earth's Atmosphere," **Journal of Atmospheric Science, 37,** 333.

Wang, W. and N. Sze, (1980), "Coupled Effects of Atmospheric N_2O and O_3 on the Earth Climate," **Nature, 286,** 589.

Weiss, R., (1980), "The Temporal and Spatial Distribution of Tropospheric Nitrous Oxide," **J. Geophys. Res., 86,** 7185.

Williams, J., (Ed.), (1978), **Carbon Dioxide, Climate and Society,** Institute for Applied Systems Analysis, Pergamon Press, New York, N.Y.

Woodwell, G., (1978), "The Carbon Dioxide Question," **Scientific American, 238,** 34.